SPINOFF

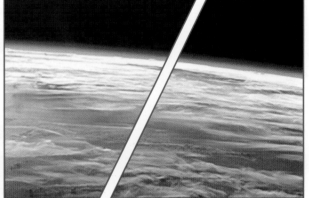

50 YEARS OF NASA-DERIVED TECHNOLOGIES (1958-2008)

Innovative Partnerships Program

2008

ISBN 978-0-16-081423-5

9 780160 814235

90000

For sale by the Superintendent of Documents, U.S. Government Printing Office
Internet: bookstore.gpo.gov   Phone: toll free (866) 512-1800;   DC area (202) 512-1800
Fax: (202) 512-2104 Mail: Stop IDCC, Washington, DC 20402-0001

ISBN 978-0-16-081423-5

Developed by
Publications and Graphics Department
NASA Center for AeroSpace Information (CASI)

On the Cover: Photographs taken from the Hubble Space Telescope and the International Space Station border a collage of past, present, and future NASA missions and spinoffs: Apollo 11 yielded emergency rescue blankets; the Space Shuttle Program improved orthotic knee joints; and research for future lunar missions produced electron beam freeform fabrication (EBF3).

Table of Contents

Denotes that R&D 100 Magazine has awarded the technology with its R&D 100 Award.

 Indicates that the Space Foundation has inducted the technology into the Space Technology Hall of Fame.

For a list of all *Spinoff* award winners since publication began in 1976, please see page 24.

Foreword

At the dawn of the space age 50 years ago, President Dwight D. Eisenhower proclaimed that "many aspects of space and space technology . . . can be helpful to all people as the United States proceeds with its peaceful program in space science and exploration. Every person has the opportunity to share through understanding in the adventures which lie ahead."

Since its founding in 1958, NASA's exploration and research missions have benefited people around the world through the expansion of our civilization's horizons, the acquisition of knowledge, and the development of new technologies and applications that provide amazing new advances in the quality of human life. The Agency also has endeavored, as President Eisenhower observed, to give every person the opportunity to share in its aeronautics and space adventures. NASA's annual *Spinoff* publication is part of that effort.

Spinoff 2008 highlights NASA's pursuit of its charter to "research, develop, verify and transfer advanced aeronautics, space, and related technologies." Among the more noteworthy NASA-derived technologies featured in this publication are:

- An advanced composite developed at NASA, now being used as insulation on thin metal wires connected to implantable cardiac resynchronization therapy (CRT) devices

- A robotic arm and hand developed for International Space Station repair that has been adapted for use in a minimally invasive knee surgery procedure, in which its precision control makes it ideal for inserting a very small implant

- A sensor originally designed for measuring fluid levels in landing gear that is now being used to measure fuel levels and detect water in the gasoline tanks of boat motors

- A welding technique perfected at NASA that is reducing time and cost in manufacturing an array of products, as well as providing enhanced design flexibility in automotive, aerospace, structural, and fluid-handling applications

While these and scores of other collateral benefits of NASA missions have enriched human society in countless ways, the true promise of the space age is just beginning. As the visionary writer Arthur C. Clarke noted, "Many, and some of the most pressing, of our terrestrial problems can be solved only by going into space. Long before it was a vanishing commodity, the wilderness as the preservation of the world was proclaimed by Thoreau. In the new wilderness of the solar system may lie the future preservation of mankind."

In this spirit, NASA is working, through research onboard the International Space Station, through robotic probes now scouting the planets, and through the development of America's next generation of human spacecraft and launch systems, to prepare for future adventures beyond Earth, beginning with our return to the Moon by 2020.

NASA's first 50 years have resulted in 12 men walking on the Moon, humanity's first up-close images of and data from the planets and their moons, and truly astounding discoveries about the nature of the universe. As a fervent believer in human progress, propelled by the dreamers and doers among us, I truly believe the best is yet to come.

Michael D. Griffin
Administrator
National Aeronautics and
Space Administration

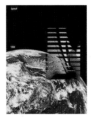

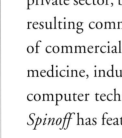

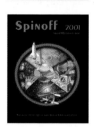

For more than 40 years, the NASA Innovative Partnerships Program has facilitated the transfer of NASA technology to the private sector, benefiting global competition and the economy. The resulting commercialization has contributed to the development of commercial products and services in the fields of health and medicine, industry, consumer goods, transportation, public safety, computer technology, and environmental resources. Since 1976, *Spinoff* has featured over 1,600 of these commercial products.

Introduction

Human curiosity has always been drawn to the wonder and mystery of the heavens. Once only the province of imagination, answers to some of the mysteries of our universe are now taking shape. The reality we witness is stunning in its beauty and humbling in its complexity and expanse. In just a half-century we have left the protective cradle of our home planet Earth, walked on another celestial body, peered into the far reaches of our universe, and sent probes into the dark and vast region of interstellar space beyond the influence of the Sun—our home star. Our minds have always reached above the clouds, but only now do we have the tools and capabilities needed to learn what is out there and explore it ourselves.

These tools and capabilities of the space age—technologies for solving seemingly impossible challenges in the harsh and unforgiving environment of space—are essential for exploring the unknown. But that is not all they can be used for. NASA also seeks Earth-bound applications for those technologies, and works with industry, universities, and other agencies to put them to use improving our everyday lives in countless ways. Finding these alternative applications, or spinoffs, for NASA-derived technology is something NASA has been doing since it was created 50 years ago. More than 1,600 examples of transferring NASA technology for public benefit have been documented in NASA's annual *Spinoff* publication since 1976. This year's 50th anniversary edition of *Spinoff* features a historical review of some of NASA's more notable spinoff successes, winning essays from an essay contest for middle-school kids describing how NASA technology affects their lives, and 50 new spinoff stories, in recognition of NASA's 50 years. Drawn from around the country, with impact around the world, these impressive new examples of how NASA technologies provide broad public benefit span modern life and lifestyle. Just a few examples include:

- Research into aerodynamics at NASA centers has been applied to improve the efficiency of tractor trailers and created a product that can improve the safety, stability, and fuel economy of numerous vehicles.

- Exploration of Mars as well as space shuttle and space station missions produced revolutionary imaging technologies now being applied to generate 360-degree views of real estate and rental properties, unprecedented panoramas of far-flung destinations, and immersive views of metropolitan areas for infrastructure monitoring and navigation.

- Technologies designed to test aircraft engine combustion chambers resulted in sensors now applied in deep well-drilling operations, where temperature and pressure increase with depth.

- Helicopter handling and stability research, long a focus of NASA's aeronautics program, has produced a stability augmentation system improving the safety of popular light helicopters often employed by police and news media.

- NASA's designs for cable-compliant joints, allowing full range of movement for rocket assemblies and robots, are now being used in a harness system for physical therapy to help people and horses recover from injuries.

We see in these technologies, and the more than 1,600 other products and processes profiled by *Spinoff* since 1976, the tangible benefits to our lives from the pursuit of sky and space. Public and private life, industry and environment, communication and transportation, healing and health, all have been profoundly affected by technologies and techniques spun off from five decades of reaching for the stars.

"Mystery creates wonder, and wonder is the basis of man's desire to understand." These memorable words from astronaut Neil Armstrong effectively summarize the drive behind NASA's pursuit of inspiration, innovation, and discovery. NASA's Innovative Partnerships Program seeks partners in this endeavour. By working together, NASA, industry, academia, other government agencies,

Douglas A. Comstock
Director
Innovative Partnerships Program

and the public can continue to push back new frontiers; increase understanding of our home planet, the universe, and our place in it; and provide broad public benefit from new applications of space-age technologies.

Aeronautics and space research continues to improve and revolutionize our lives with tangible and remarkable benefits for all. The legacy of public benefit from NASA technologies and the new examples highlighted here in *Spinoff* 2008 are the direct result of the U.S. Government's vibrant civil space and aeronautics program, dedicated to active and productive collaboration with private industry and academia. May the next 50 years prove as exciting, enlightening, and rewarding as NASA's first 50 years.

50 Years of NASA-Derived Technologies

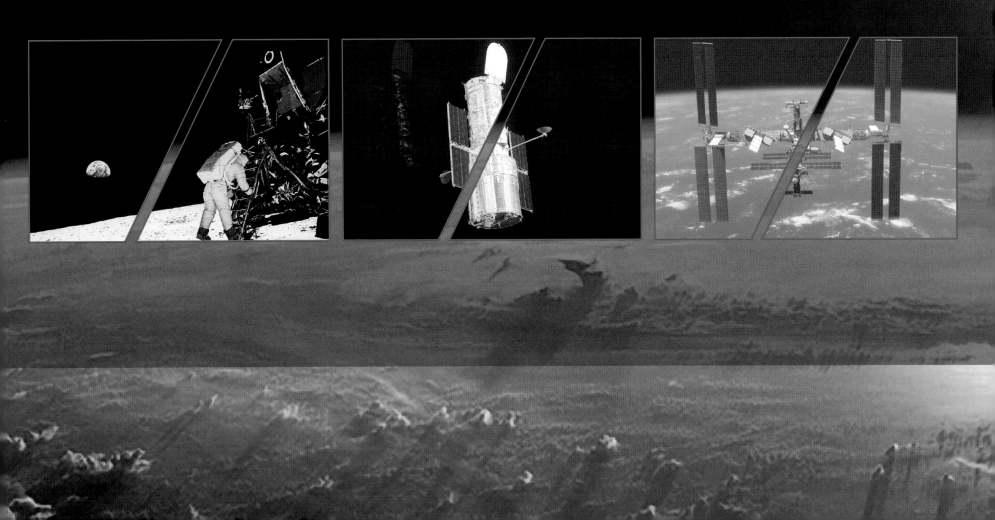

Fifty years ago, the U.S. Congress passed the National Aeronautics and Space Act of 1958, forming NASA and calling for the widespread dissemination of its newfound technologies to the public. Five decades of daring to challenge the impossible and brave new frontiers of exploration and technology has brought NASA, and indeed humankind, countless discoveries, revelations, and dramatic moments of pride and wonder.

Inspiration fueling innovation, innovation enabling discovery; this is NASA's legacy and future, and the sometimes serendipitous mechanism through which the lives of the American public, and people worldwide, have benefited. Exploring the cosmos has revolutionized medicine, transportation, public safety, recreation, environmental monitoring and resource management, computer technology, industrial productivity, and our perception of the planet on which we live and the universe of which our Earth is one small part.

50 Years of Exploration and Innovation

NASA's 50th anniversary is an opportunity to celebrate the power of inspiration, innovation, and discovery. While looking forward to the bright promise held by the future of space exploration, an eye toward the past highlights just how far technology and understanding have come in the last five decades, amid milestone moments both poignant and profound as the first steps were taken into the vastness beyond Earth's atmosphere.

This timeline collects the most memorable missions that brought our fellow men and women into the company of the stars, as well as research into the cutting edge of aeronautics design and practices. A selection of notable benefits realized by society follows each entry, highlighting how the pursuit of sky and stars continually revolutionizes our daily lives on Earth. The entries are listed by program start date, to frame the context of the research and development, and connect to the time scale at the date of first launch. As the last page turns from this proud and rich history to look at the missions yet to come, our imaginations are free to consider what marvelous developments these technologies may bring to our home and life.

We set sail on this new sea because there is new knowledge to be gained, and new rights to be won, and they must be won and used for the progress of all people.

-President John F. Kennedy, 1962

NASA 50th Anniversary Essay Competition

P art of NASA's mission is to inspire the next generation of engineers, scientists, and explorers. NASA's Innovative Partnerships Program, in conjunction with the Office of Education, conducted the NASA 50th Anniversary Essay Competition during academic year 2007-2008 to inspire and encourage middle school students to continue with science, technology, engineering, and mathematics.

First, students described benefits in everyday life from technologies built by NASA over the last 50 years; the winning essay is published below. Second, NASA challenged students to imagine how their everyday lives will be changed by NASA technology years into the future; that essay is found at the end of the timeline.

First Prize: $5,000 college scholarship and four VIP trips to the Kennedy Space Center to watch the STS-125 space shuttle launch
Jackson Warley, 7th Grade, Renaissance Academy in Colorado Springs, Colorado
Teacher: Mr. Ron Hamilton

Describe how you benefit today in everyday life due to NASA aerospace technology and spinoffs from the last 50 years.

Since NASA's inception fifty years ago, the technology it has developed has changed everyday life dramatically. NASA technologies adapted for use outside the space program include smoke detectors with adjustable sensitivity, memory foam, space blankets, advanced microencapsulation technology, foamless toothpaste, space pens, advanced heat-resistant composites and other inventions. However, it is not only NASA technology that has been beneficial to us all; it is also the underlying spirit and principles of NASA.

It seems that to some people it is hard to think of anything used in everyday life that was developed by the space program. In fact, we are surrounded by these objects. You may have seen pens that write at all angles in stores. These were developed for use in space. Have you ever been to a camping store? If so, you will have seen reflective space blankets. These (as their name would suggest) were also developed for the space program. If you play football you may have used a shock absorbing helmet. These make use of memory foam (also known as temper foam); a material developed by NASA for use in aircraft seats. You might also have slept in a bed made of this material. If you have ever watched television transmitted from a satellite dish, you have made use of another NASA inspired technology. Although NASA did not technically invent television satellite dishes, it has developed ways to clarify signals hence reducing signal noise.

Just because there is NASA-developed technology used for non-aerospace applications doesn't mean that you use it in your day-to-day life. Some NASA technology is used more frequently for industrial applications.

This includes a noteworthy invention that is used to clean up oil spills. It consists of microcapsules of beeswax that stimulate the natural microbes to consume the oil. This makes use of NASA microencapsulation technology developed at the Jet Propulsion Laboratory in California. Not only does this product stimulate microbial life to remove the oil, the beeswax spheres actually absorb oil into their hollow centers and dispose of it safely. It is important that the oil is not allowed to settle, another reason why this technology is so effective. This is just one example of NASA inspired technology that benefits our daily lives indirectly by improving the condition of the overall environment.

The final thing that benefits us in everyday life isn't even a piece of technology. This is, quite simply put, the overall spirit of NASA. NASA heeds the basic human calling to explore the unknown and in doing so, gives people motivation, albeit not through advanced technology. One hundred years ago, before NASA was founded, the idea of humans in space would have seemed impossible. Now, it is commonplace. Through daring and determination, even the wildest dreams can be realized. "Failure is not an option." Never give up, even under the most grim circumstances. These are the concepts through which NASA has inspired us over the fifty years that it has existed.

50 Years of Exploration and Innovation

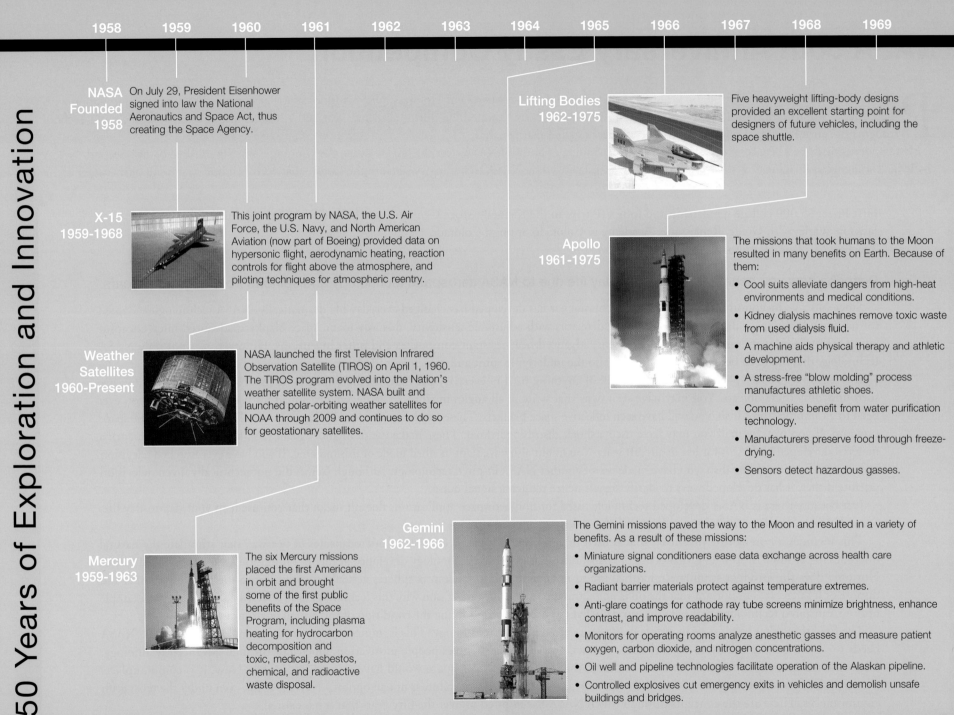

1958　1959　1960　1961　1962　1963　1964　1965　1966　1967　1968　1969

NASA Founded 1958

On July 29, President Eisenhower signed into law the National Aeronautics and Space Act, thus creating the Space Agency.

X-15 1959-1968

This joint program by NASA, the U.S. Air Force, the U.S. Navy, and North American Aviation (now part of Boeing) provided data on hypersonic flight, aerodynamic heating, reaction controls for flight above the atmosphere, and piloting techniques for atmospheric reentry.

Weather Satellites 1960-Present

NASA launched the first Television Infrared Observation Satellite (TIROS) on April 1, 1960. The TIROS program evolved into the Nation's weather satellite system. NASA built and launched polar-orbiting weather satellites for NOAA through 2009 and continues to do so for geostationary satellites.

Mercury 1959-1963

The six Mercury missions placed the first Americans in orbit and brought some of the first public benefits of the Space Program, including plasma heating for hydrocarbon decomposition and toxic, medical, asbestos, chemical, and radioactive waste disposal.

Lifting Bodies 1962-1975

Five heavyweight lifting-body designs provided an excellent starting point for designers of future vehicles, including the space shuttle.

Apollo 1961-1975

The missions that took humans to the Moon resulted in many benefits on Earth. Because of them:

- Cool suits alleviate dangers from high-heat environments and medical conditions.
- Kidney dialysis machines remove toxic waste from used dialysis fluid.
- A machine aids physical therapy and athletic development.
- A stress-free "blow molding" process manufactures athletic shoes.
- Communities benefit from water purification technology.
- Manufacturers preserve food through freeze-drying.
- Sensors detect hazardous gasses.

Gemini 1962-1966

The Gemini missions paved the way to the Moon and resulted in a variety of benefits. As a result of these missions:

- Miniature signal conditioners ease data exchange across health care organizations.
- Radiant barrier materials protect against temperature extremes.
- Anti-glare coatings for cathode ray tube screens minimize brightness, enhance contrast, and improve readability.
- Monitors for operating rooms analyze anesthetic gasses and measure patient oxygen, carbon dioxide, and nitrogen concentrations.
- Oil well and pipeline technologies facilitate operation of the Alaskan pipeline.
- Controlled explosives cut emergency exits in vehicles and demolish unsafe buildings and bridges.

This timeline presents a few dozen of the more than 1,600 technologies profiled by *Spinoff* over the last 30+ years.

F-8C Digital Fly-By-Wire Control System 1972-1985

Now widely used by commercial airliners, the digital fly-by-wire control system improved maneuver control, ride, and combat survivability for military aircraft.

Winglets 1979-1981

Winglets applied to the tips of the main wings on KC-135 aircraft improved vehicle aerodynamics and fuel efficiency and are now universally accepted.

Landsat 1-7 1966-Present

For 34 years, the Landsat series of satellite missions jointly managed by NASA and the U.S. Geological Survey has collected information about Earth from space. Landsat satellites have taken specialized digital photographs of Earth's continents and surrounding coastal regions, enabling people to study many aspects of our planet and evaluate the dynamic changes caused by natural processes and human practices. This unparalleled data archive gives scientists the ability to assess changes in Earth's landscape and atmosphere.

Voyager 1977-Present

30 years after launch, Voyager 1 and 2 are entering interstellar space. Both spacecraft still send information through the Deep Space Network, and have brought other benefits:

- AIDS research and e-commerce utilize software to identify data deviations.
- Astronomers and space enthusiasts employ software offering views recorded by spacecraft.

Space Shuttle 1969-Present

The Space Shuttle Program alone has generated more than 100 technology spinoffs. As a result of shuttle research:

- Miniaturized heart pumps save lives.
- Thermal protection system materials protect racecar drivers.
- Bioreactors help chemists design new therapeutic drugs and antibodies.
- Compact laboratory instruments allow faster blood analysis.
- Sensitive hand-held infrared cameras scan for forest fires.
- Rocket fuel helps destroy land mines.
- Light-emitting diodes treat cancerous tumors.
- Prosthetic limbs are lighter and stronger.
- An extrication tool removes accident victims from wrecked vehicles.
- Municipalities track and reassign emergency and public works vehicles.
- Law enforcement agencies can improve the resolution of crime scene video.

Skylab 1973-1979

America's first space station and orbital science and engineering laboratory provided an unprecedented platform for experimentation and returned myriad benefits:

- Hand-held ultrasonic systems detect indications of bearing failure.
- Monitoring systems evaluate particulate matter in gas streams.
- Computerized solar water heaters save energy.
- Cryogenic liquid methane tanks store aircraft fuel.
- Emergency and night lighting systems save energy.
- Solar screens cut 70 to 80 percent of heat and glare.
- New wire-and-rod grounding systems prevent corrosion.
- Negative pressure techniques relieve respiratory distress in infants.

1983	1984	1985	1986	1987	1988	1989	1990	1991	1992	1993	1994	1995

X-29 Flight Research 1981-1990

X-29 flight research aircraft demonstrated forward-swept wing technology and provided data on aeroelastic tailoring, active controls, and canard effects.

Hubble Space Telescope 1990-Present

Perhaps the world's most famous telescope, Hubble has given us more than close-up views of our galaxy:

- Surgeons perform micro-invasive arthroscopic surgery with increased precision.
- Precision optics and advanced scheduling software optimizes semiconductor manufacturing.
- Software allows astronomers to locate, identify, and acquire images of deep sky objects.
- Imaging technology makes breast biopsies less invasive and more accurate.

NASA/FAA Wind Shear Program 1986-1993

A joint study by NASA and the Federal Aviation Administration on the cause of wind shear resulted in a better understanding of corrective actions, outcome procedures, and technologies.

NASA/FAA Research into Efficient Airspace Operations 1994-2001

The Terminal Area Productivity (TAP) Program sought to achieve clear-weather capacity, safely and affordably, in instrument-weather conditions. Benefits include:

- Mathematically modeled wake vortices to determine safe aircraft separation standards.
- A new computer system to assist flight controllers, called the Aircraft Vortex Spacing System.
- A three-dimensional auditory display system for ground operations, including a computer-generated voice that provides verbal warnings of impending collisions with other aircraft or vehicles.

International Space Station
1993-Present

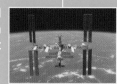

Still under construction and orbiting 200 miles above, the International Space Station is yielding benefits here on Earth:

- A novel, quick-fastening nut for use in firefighting, aerospace, gas fittings, and manufacturing.
- Hand-held devices warn pilots of dangerous or deteriorating cabin pressure.
- An air purifier kills 93.3 percent of airborne pathogens, including anthrax.
- Robotic arms assist in human-collaborative medical surgery and emergency response to chemical, biological, and nuclear material spills.
- Reverse osmosis technology is used to clean the water runoff from landfills.
- Superelastic and high-damping golf clubs.
- A video headset offers people with low vision a view of their surroundings.
- 360° immersive digital representations provide consumers with views of the latest automobiles, hotel accommodations, and real estate.

Environmental Research Aircraft and Sensor Technology (ERAST)
1994-2003

The remotely piloted, solar-powered "Helios" vehicle flew to the record-breaking altitude of 96,863 feet, leading the way for future high-altitude, long-duration, solar-powered aircraft.

Earth Observing System
1999-Present

NASA launched its first series of EOS satellites from 1999 through 2004, and continues to upgrade and enhance this constellation of satellites to detect and measure global change. Observations from these satellites provide much of the basis of our understanding of climate change, and find myriad applications in our economy and society.

NASA's 50th Anniversary
2008

This year, America's Space Agency celebrates 50 years of innovation, inspiration, and discovery, enabling us to learn more about ourselves, our world, and how to manage and protect it.

Mars Exploration Rovers
1998-Present

Using the accumulated technology and knowledge of over 30 years of Mars exploration programs, these twin vehicles, launched in June (Spirit) and July (Opportunity) 2003, continue their pursuit of geological clues about current and past environmental conditions on Mars. Benefits include:

- A robot that can climb grades up to 60 percent and survive submersion in water up to 6.6 feet deep, and possesses flippers that propel it over obstructions and through rocks, rubble, and debris. It is being used during combat operations to provide soldiers with a safe first look at potentially hostile environments.
- A mineral identification tool now helping U.S. law enforcement agencies and military personnel identify suspicious liquid and solid substances.
- Panoramic imaging and photographic systems being used to photograph and map cities, towns, and landscapes for municipal, commercial, and personal applications.

Vision for Space Exploration
2004

On January 14, President Bush unveiled the Vision for Space Exploration, laying out plans for NASA to complete assembly of the International Space Station by 2010, retire the space shuttle, and create new launch vehicles and programs that will return NASA to the Moon and then on to Mars.

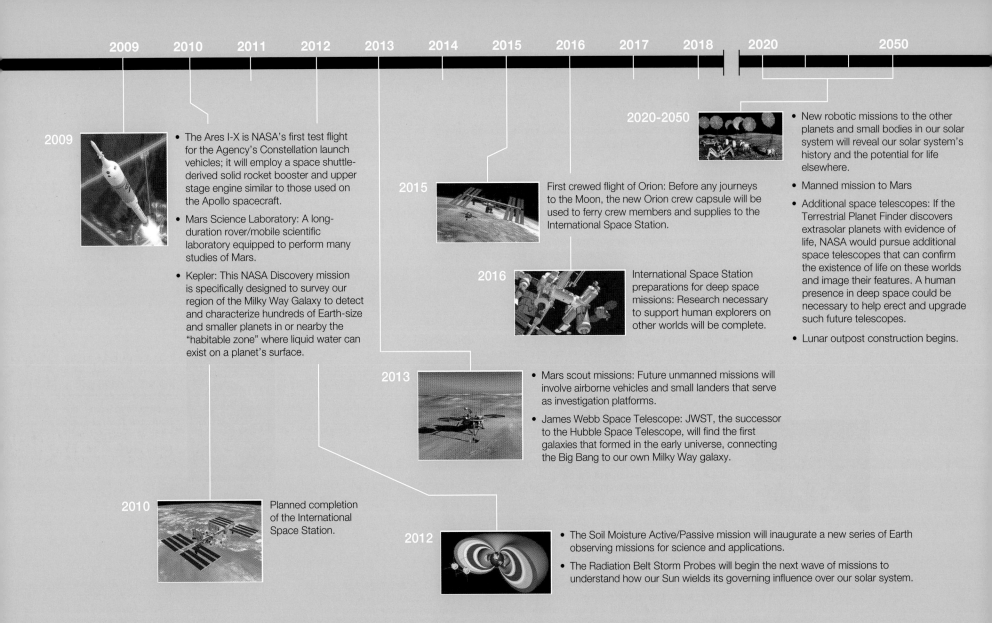

2009 · The Ares I-X is NASA's first test flight for the Agency's Constellation launch vehicles; it will employ a space shuttle-derived solid rocket booster and upper stage engine similar to those used on the Apollo spacecraft.

· Mars Science Laboratory: A long-duration rover/mobile scientific laboratory equipped to perform many studies of Mars.

· Kepler: This NASA Discovery mission is specifically designed to survey our region of the Milky Way Galaxy to detect and characterize hundreds of Earth-size and smaller planets in or nearby the "habitable zone" where liquid water can exist on a planet's surface.

2015 First crewed flight of Orion: Before any journeys to the Moon, the new Orion crew capsule will be used to ferry crew members and supplies to the International Space Station.

2016 International Space Station preparations for deep space missions: Research necessary to support human explorers on other worlds will be complete.

2020-2050

· New robotic missions to the other planets and small bodies in our solar system will reveal our solar system's history and the potential for life elsewhere.

· Manned mission to Mars

· Additional space telescopes: If the Terrestrial Planet Finder discovers extrasolar planets with evidence of life, NASA would pursue additional space telescopes that can confirm the existence of life on these worlds and image their features. A human presence in deep space could be necessary to help erect and upgrade such future telescopes.

· Lunar outpost construction begins.

2013 · Mars scout missions: Future unmanned missions will involve airborne vehicles and small landers that serve as investigation platforms.

· James Webb Space Telescope: JWST, the successor to the Hubble Space Telescope, will find the first galaxies that formed in the early universe, connecting the Big Bang to our own Milky Way galaxy.

2010 Planned completion of the International Space Station.

2012 · The Soil Moisture Active/Passive mission will inaugurate a new series of Earth observing missions for science and applications.

· The Radiation Belt Storm Probes will begin the next wave of missions to understand how our Sun wields its governing influence over our solar system.

Space is for everybody. It's not just for a few people in science or math, or for a select group of astronauts. That's our new frontier out there, and it's everybody's business to know about space.

-Christa McAuliffe, 1985

As NASA celebrates 50 years of scientific and technological excellence that have powered us into the 21st century, it reflects on signature accomplishments that are enduring icons of human achievement. Among those accomplishments are technological innovations and scientific discoveries that have improved and shaped our lives on Earth in a myriad of ways. In looking forward to a promising new era of inspiration, innovation, and discovery, the second topic in the 50th Anniversary Essay Contest required students to imagine how their everyday lives will have changed because of NASA aerospace technology years into the future.

Second Prize: $2,500 college scholarship
Grace Nowadly, 7th Grade, Berkeley Middle School in Williamsburg, Virginia
Teacher: Ms. Kathy Poe

Describe how, 50 years from now, your everyday life may benefit from NASA's future aerospace technology.

"Hurry up, Grace! You're going to miss it!"

"Be there in a minute!" I called back to my husband. I grew more impatient by the second as, one by one, my popcorn kernels began to pop. The microwave finally started to beep and I grabbed the popcorn and gave it to my granddaughter who ran back to the room in which the rest of her family was sitting. I slowly walked back and sat down just in time to see a hazy image of a spacecraft slowly moving towards a small, red planet.

"I never thought I'd live to see the day when astronauts landed on Mars. My mother used to tell me about her whole family watching the landing on the moon, and now my whole family is watching the landing on Mars! Just think about how far we've come. First the moon, now Mars! What's next?" I said out loud as my grandchildren watched the screen, only half listening.

Hmm…What is next? I thought to myself. I looked over at my son. He was an air traffic controller at a local airport. *Oh yeah! He told me that NASA was going to set up a new program that makes air traffic control easier.* I turned back to the television and watched the ship come closer and closer to the tiny planet. I was watching the television, but I was doing lots of thinking.

I also heard on the news somewhere that NASA will launch a satellite that can track the spreading of diseases, I thought to myself. *That could help doctors and scientists to control pandemics. If they knew where diseases were rapidly spreading, they could evacuate healthy people from the area, help the infected people, and keep people from coming in the area.*

I looked around the room and saw my three grandchildren sitting around the television in big bean bag chairs. *Will NASA try to help them somehow?* I remembered when I brought them to the Air and Space Museum in Washington D.C. They looked at every brochure of educational programs available to them. There were camps, classes, and other programs that all focused on NASA and space exploration.

I looked at the television and heard rejoicing through out the room as the spacecraft landed on the smoky, red, surface of Mars. The kids jumped up and down and ran around the house. Everyone was happy and watched eagerly for the astronauts' next move. As they slowly stepped out of the tiny door onto a planet never visited by man, I couldn't help but think, *Well, NASA will keep making everyday life better, one step at a time.*

NASA Technologies Benefit Our Lives

Space exploration has created new markets and new technologies that have spurred our economy and changed our lives in many ways. This year, NASA unveiled two new complementary interactive Web features, NASA City and NASA @ Home, available at www.nasa.gov/city. The new features highlight how space pervades our lives, invisible yet critical to so many aspects of our daily activities and well-being.

Health and Medicine

Light-Emitting Diodes (LEDs)

Red light-emitting diodes are growing plants in space and healing humans on Earth. The LED technology used in NASA space shuttle plant growth experiments has contributed to the development of medical devices such as award-winning WARP 10, a hand-held, high-intensity, LED unit developed by Quantum Devices Inc. The WARP 10 is intended for the temporary relief of minor muscle and joint pain, arthritis, stiffness, and muscle spasms, and also promotes muscle relaxation and increases local blood circulation. The WARP 10 is being used by the U.S. Department of Defense and U.S. Navy as a noninvasive "soldier self-care" device that aids front-line forces with first aid for minor injuries and pain, thereby improving endurance in combat. The next-generation WARP 75 has been used to relieve pain in bone marrow transplant patients, and will be used to combat the symptoms of bone atrophy, multiple sclerosis, diabetic complications, Parkinson's disease, and in a variety of ocular applications. (*Spinoff* 2005, 2008)

Infrared Ear Thermometers

Diatek Corporation and NASA developed an aural thermometer, which weighs only 8 ounces and uses infrared astronomy technology to measure the amount of energy emitted by the eardrum, the same way the temperature of stars and planets is measured. This method avoids contact with mucous membranes, virtually eliminating the possibility of cross infection, and permits rapid temperature measurement of newborn, critically-ill, or incapacitated patients. NASA supported the Diatek Corporation, a world leader in electronic thermometry, through the Technology Affiliates Program. (*Spinoff* 1991)

Artificial Limbs

NASA's continued funding, coupled with its collective innovations in robotics and shock-absorption/comfort materials are inspiring and enabling the private sector to create new and better solutions for animal and human prostheses. Advancements such as Environmental Robots Inc.'s development of artificial muscle systems with robotic sensing and actuation capabilities for use in NASA space robotic and extravehicular activities are being adapted to create more functionally dynamic artificial limbs (*Spinoff* 2004). Additionally, other private-sector adaptations of NASA's temper foam technology have brought about custom-mold-able materials offering the natural look and feel of flesh, as well as preventing friction between the skin and the prosthesis, and heat/moisture buildup. (*Spinoff* 2005)

Ventricular Assist Device

Collaboration between NASA, Dr. Michael DeBakey, Dr. George Noon, and MicroMed Technology Inc. resulted in a lifesaving heart pump for patients awaiting heart transplants. The MicroMed DeBakey ventricular assist device (VAD) functions as a "bridge to heart transplant" by pumping blood throughout the body to keep critically ill patients alive until a donor heart is available. Weighing less than 4 ounces and measuring 1 by 3 inches, the pump is approximately one-tenth the size of other currently marketed pulsatile VADs. This makes it less invasive and ideal for smaller adults and children. Because of the pump's small size, less than 5 percent of the patients implanted developed device-related infections. It can operate up to 8 hours on batteries, giving patients the mobility to do normal, everyday activities. (*Spinoff* 2002)

Transportation

Anti-Icing Systems

NASA funding under the Small Business Innovation Research (SBIR) program and work with NASA scientists advanced the development of the certification and integration of a thermoelectric deicing system called Thermawing, a DC-powered air conditioner for single-engine aircraft called Thermacool, and high-output alterna-tors to run them both. Thermawing, a reliable anti-icing and deicing system, allows pilots to safely fly through ice encounters and provides pilots of

single-engine aircraft the heated wing technology usually reserved for larger, jet-powered craft. Thermacool, an innovative electric air conditioning system, uses a new compressor whose rotary pump design runs off an energy-efficient, brushless DC motor and allows pilots to use the air conditioner before the engine even starts. (*Spinoff* 2007)

Highway Safety

Safety grooving, the cutting of grooves in concrete to increase traction and prevent injury, was first developed to reduce aircraft accidents on wet runways. Represented by the International Grooving and Grinding Association, the industry expanded into highway and pedestrian applications. The technique originated at Langley Research Center, which assisted in testing the grooving at airports and on highways. Skidding was reduced, stopping distance decreased, and a vehicle's cornering ability on curves was increased. The process has been extended to animal holding pens, steps, parking lots, and other potentially slippery surfaces. (*Spinoff* 1985)

Improved Radial Tires

Goodyear Tire and Rubber Company developed a fibrous material, five times stronger than steel, for NASA to use in parachute shrouds to soft-land the Vikings on the Martian surface. The fiber's chain-like molecular structure gave it incredible strength in proportion to its weight. Recognizing the increased strength and durability of the material, Goodyear expanded the technology and went on to

produce a new radial tire with a tread life expected to be 10,000 miles greater than conventional radials. (*Spinoff* 1976)

Chemical Detection

NASA contracted with Intelligent Optical Systems (IOS) to develop moisture- and pH-sensitive sensors to warn of potentially dangerous corrosive conditions in aircraft before significant structural damage occurs. This new type of sensor, using a specially manufactured optical fiber whose entire length is chemically sensitive, changes color in response to contact with its target. After completing the work with NASA, IOS was tasked by the U.S. Department of Defense to further develop the sensors for detecting chemical warfare agents and potential threats, such as toxic industrial compounds and nerve agents, for which they proved just as successful. IOS has additionally sold the chemically sensitive fiber optic cables to major automotive and aerospace companies, who are finding a variety of uses for the devices such as aiding experimentation with nontraditional power sources, and as an economical "alarm system" for detecting chemical release in large facilities. (*Spinoff* 2007)

Public Safety

Video Enhancing and Analysis Systems

Intergraph Government Solutions developed its Video Analyst System (VAS) by building on Video Image Stabilization and Registration (VISAR) technology created by NASA to help FBI agents analyze video footage. Originally used for enhancing video images from nighttime videotapes made with hand-held camcorders, VAS is a state-of-the-art, simple, effective, and affordable tool for video enhancement and analysis

offering benefits such as support of full-resolution digital video, stabilization, frame-by-frame analysis, conversion of analog video to digital storage formats, and increased visibility of filmed subjects without altering underlying footage. Aside from law enforcement and security applications, VAS has also been adapted to serve the military for reconnaissance, weapons deployment, damage assessment, training, and mission debriefing. (*Spinoff* 2001)

Land Mine Removal

Due to arrangements such as the one between Thiokol Propulsion and NASA that permits Thiokol to use NASA's surplus rocket fuel to produce a flare that can safely destroy land mines, NASA is able to reduce propellant waste without negatively impacting the environment, and Thiokol is able to access the materials needed to develop the Demining Device flare. The Demining Device flare uses a battery-triggered electric match to ignite and neutralize land mines in the field without detonation. The flare uses solid rocket fuel to burn a hole in the mine's case and burn away the explosive contents so the mine can be disarmed without hazard. (*Spinoff* 2000)

Fire-Resistant Reinforcement

Built and designed by Avco Corporation, the Apollo heat shield was coated with a material whose purpose was to burn and thus dissipate energy during reentry while charring, to form a protective coating to block heat penetration. NASA subsequently funded Avco's development of other applications of the heat shield, such as fire-retardant paints and foams for aircraft, which led to the world's first intumescent epoxy material, which expands in volume when exposed to heat or flames, acting as an insulating barrier and dissipating heat through burn-off. Further innovations based on this product include steel coatings devised to make high-rise buildings and public structures safe by swelling to provide a tough and stable insulating layer over the steel for up to 4 hours of fire protection, ultimately to slow building collapse and provide more time for escape. (*Spinoff* 2006)

Firefighter Gear

Firefighting equipment widely used throughout the United States is based on a NASA development that coupled Agency design expertise with lightweight materials developed for the U.S. Space Program. A project that linked NASA and the National Bureau of Standards resulted in a lightweight breathing system including face mask, frame, harness, and air bottle, using an aluminum composite material developed by NASA for use on rocket casings. Aerospace technology has been beneficially transferred to civil-use applications for years, but perhaps the broadest fire-related technology transfer is the breathing apparatus worn by firefighters for protection from smoke inhalation injury. Additionally, radio communications are essential during a fire to coordinate

hose lines, rescue victims, and otherwise increase efficiency and safety. NASA's inductorless electronic circuit technology contributed to the development of a lower-cost, more rugged, short-range two-way radio now used by firefighters. NASA also helped develop a specialized mask weighing less than 3 ounces to protect the physically impaired from injuries to the face and head, as well as flexible, heat-resistant materials—developed to protect the space shuttle on reentry—which are being used both by the military and commercially in suits for municipal and aircraft-rescue firefighters. (*Spinoff* 1976)

Consumer, Home, and Recreation

Temper Foam

As the result of a program designed to develop a padding concept to improve crash protection for airplane passengers, Ames Research Center developed a foam material with unusual properties. The material is widely used and commonly known as temper foam or "memory foam." The material has been incorporated into a host of widely used and recognized products including mattresses, pillows, military and civilian aircraft, automobiles and motorcycles, sports safety equipment, amusement park rides and arenas, horseback saddles, archery targets, furniture, and human and animal prostheses. Its high-energy absorption and soft characteristics not only offer superior protection in the event of an accident or impact, but enhanced comfort and support for passengers on long flights or those seeking restful sleep. Today, temper foam is being employed by NASCAR to provide added safety in racecars. (*Spinoff* 1976, 1977, 1979, 1988, 1995, 2002, 2005)

Enriched Baby Food

Commercially available infant formulas now contain a nutritional enrichment ingredient that traces its existence to NASA-sponsored research that explored the potential of algae as a recycling agent for long-duration space travel. The substance, formulated into the products life'sDHA and life'sARA, can be found in over 90 percent of the infant formulas sold in the United States, and are added to the infant formulas sold in over 65 additional countries. The products were developed and are manufactured by Martek Biosciences Corporation, which has pioneered the commercial development of products based on microalgae; the company's founders and principal scientists acquired their expertise in this area while working on the NASA program. (*Spinoff* 1996, 2008)

Portable Cordless Vacuums

Apollo and Gemini space mission technologies created by Black & Decker have helped change the way we clean around the house. For the Apollo space mission, NASA required a portable, self-contained drill capable of extracting core samples from below the lunar surface. Black & Decker was tasked with the job, and developed a computer program to optimize the design of the drill's motor and insure minimal power consumption. That computer program led to the development of a cordless miniature vacuum cleaner called the Dustbuster. (*Spinoff* 1981)

Freeze Drying Technology

In planning for the long-duration Apollo missions, NASA conducted extensive research into space food. One of the techniques developed was freeze drying—Action Products commercialized this technique, concentrating on snack food. The foods are cooked, quickly frozen, and then slowly heated in a vacuum chamber to remove the ice crystals formed by the freezing process. The final product retains 98 percent of its nutrition and weighs only 20 percent of its original weight. Today, one of the benefits of this advancement in food preparation includes simple nutritious meals available to handicapped and otherwise homebound senior adults unable to take advantage of existing meal programs sponsored by government and private organizations. (*Spinoff* 1976, 1994)

Environmental and Agricultural Resources

Harnessing Solar Energy

Homes across the country are now being outfitted with modern, high-performance, low-cost, single crystal silicon solar power cells that allow them to reduce their traditional energy expenditures and contribute to pollution reduction. The advanced technology behind these solar devices—which are competitively-priced and provide up to 50 percent more power than conventional solar cells—originated with the efforts of a NASA-sponsored 28-member coalition of companies, government groups, universities, and nonprofits forming the Environmental Research Aircraft and Sensor Technology (ERAST) Alliance. ERAST's goal was to foster the development of remotely piloted aircraft intended to fly unmanned at high altitudes for days at a time, requiring advanced solar power sources that did not add weight. As a result, SunPower Corporation created the most advanced silicon-based cells available for terrestrial or airborne applications. (*Spinoff* 2005)

Pollution Remediation

A product using NASA's microencapsulating technology is available to consumers and industry enabling them to safely and permanently clean petroleum-based pollutants from water. The microencapsulated wonder, Petroleum Remediation Product or "PRP," has revolutionized the way oil spills are cleaned. The basic technology behind PRP is thousands of microcapsules—tiny balls of beeswax with hollow centers. Water cannot penetrate the microcapsule's cell, but oil is absorbed right into the beeswax spheres as they float on the water's surface. Contaminating chemical compounds that originally come from crude oil (such as fuels, motor oils, or petroleum hydrocarbons) are caught before they settle, limiting damage to ocean beds. (*Spinoff* 1994, 2006)

Water Purification

NASA engineers are collaborating with qualified companies to develop a complex system of devices intended to sustain the astronauts living on the International Space Station and, in the future, those who go on to explore the Moon. This system, tentatively scheduled for launch in 2008, will make use of available resources by turning wastewater from respiration, sweat, and urine into drinkable water. Commercially, this system is benefiting people all over the world who need affordable, clean water. By combining the benefits of chemical adsorption, ion exchange, and ultra-filtration processes, products using this technology yield safe, drinkable water from the most challenging sources, such as in underdeveloped regions where well water may be heavily contaminated. (*Spinoff* 1995, 2006)

Computer Technology

Better Software

From real-time weather visualization and forecasting, high-resolution 3-D maps of the Moon and Mars, to real-time tracking of the International Space Station and the space shuttle, NASA is collaborating with Google Inc. to solve a variety of challenging technical problems ranging from large-scale data management and massively distributed computing, to human-computer interfaces—with the ultimate goal of making the vast, scattered ocean of data more accessible and usable. With companies like InterSense, NASA continues to fund and collaborate on other software advancement initiatives benefiting such areas as photo/video image enhancement, virtual-reality/design, simulation training, and medical applications. (*Spinoff* 2005)

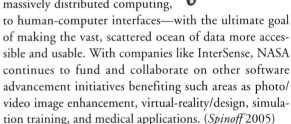

Structural Analysis

NASA software engineers have created thousands of computer programs over the decades equipped to design, test, and analyze stress, vibration, and acoustical properties of a broad assortment of aerospace parts and structures (before prototyping even begins). The NASA Structural Analysis Program, or NASTRAN, is considered one of the most successful and widely-used NASA software programs. It has been used to design everything from Cadillacs to roller coaster rides. Originally created for spacecraft design, NASTRAN has been employed in a host of non-aerospace applications and is available to industry through NASA's Computer Software Management and Information Center (COSMIC). COSMIC maintains a library of computer programs from NASA and other government agencies and offers them for sale at a fraction of the cost of developing a new program, benefiting companies around the world seeking to solve the largest, most difficult engineering problems. (*Spinoff* 1976, 1977, 1978, 1979, 1980, 1981, 1982, 1986, 1988, 1990, 1991, 1998)

Refrigerated Internet-Connected Wall Ovens

Embedded Web Technology (EWT) software—originally developed by NASA for use by astronauts operating experiments on available laptops from anywhere on the International Space Station—lets a user monitor and/or control a device remotely over the Internet. NASA supplied this technology and guidance to TMIO LLC, who went on to develop a low-cost, real-time remote control and monitoring of a new intelligent oven product named "ConnectIo." With

combined cooling and heating capabilities, ConnectIo provides the convenience of being able to store cold food where it will remain properly refrigerated until a customized pre-programmable cooking cycle begins. The menu allows the user to simply enter the dinner time, and the oven automatically switches from refrigeration to the cooking cycle, so that the meal will be ready as the family arrives home for dinner. (*Spinoff* 2005)

Industrial Productivity

Powdered Lubricants

NASA's scientists developed a solid lubricant coating material that is saving the manufacturing industry millions of dollars. Developed as a shaft coating to be deposited by thermal spraying to protect foil air bearings used in oil-free turbomachinery, like gas turbines, this advanced coating, PS300, was meant to be part of a larger project: an oil-free aircraft engine capable of operating at high temperatures with increased reliability, lowered weight, reduced maintenance, and increased power. PS300 improves efficiency, lowers friction, reduces emissions, and has been used by NASA in advanced aeropropulsion engines, refrigeration compressors, turbochargers, and hybrid electrical turbogenerators. ADMA Products has found widespread industrial applications for the material. (*Spinoff* 2005)

Improved Mine Safety

An ultrasonic bolt elongation monitor developed by a NASA scientist for testing tension and high-pressure loads on bolts and fasteners has continued to evolve over the past three decades. Today, the same scientist and Luna Innovations are using a digital adaptation of this same device for a plethora of different applications, including non-destructive evaluation of railroad ties, groundwater

analysis, radiation dosimetry, and as a medical testing device to assess levels of internal swelling and pressure for patients suffering from intracranial pressure and compartment syndrome, a painful condition that results when pressure within muscles builds to dangerous levels. The applications for this device continue to expand. (*Spinoff* 1978, 2005, 2008)

Food Safety Systems

Faced with the problem of how and what to feed an astronaut in a sealed capsule under weightless conditions while planning for human space flight, NASA enlisted the aid of The Pillsbury Company to address two principal concerns: eliminating crumbs of food that might contaminate the spacecraft's atmosphere and sensitive instruments, and assuring absolute freedom from potentially catastrophic disease-producing bacteria and toxins. Pillsbury developed the Hazard Analysis and Critical Control Point (HACCP) concept, potentially one of the most far-reaching space spinoffs, to address NASA's second concern. HACCP is designed to prevent food safety problems rather than to catch them after they have occurred. The U.S. Food and Drug Administration has applied HACCP guidelines for the handling of seafood, juice, and dairy products. (*Spinoff* 1991)

NASA Technology Award Winners

Since its inception in 1976, *Spinoff* has featured myriad award-winning technologies that have been recognized by NASA and industry as forerunners in innovation. Here is a chronology of these winners, including the year(s) they were featured in *Spinoff* and the year they were awarded one (or more) of the following:

R&D 100
The R&D 100 Awards were established in 1963 to pick the 100 most technologically significant new products invented each year. For 45 years, the prestigious R&D 100 Awards have been helping provide new products with the needed recognition for success in the marketplace. Winning an R&D 100 Award provides a mark of excellence known to industry, government, and academia.

Space Technology Hall of Fame
The Space Foundation and NASA created the Space Technology Hall of Fame in 1988. The award recognizes the life-changing technologies emerging from America's space programs; honors the scientists, engineers, and innovators responsible; and communicates to the American public the significance of these technologies as a return on investment in their Space Program.

NASA Invention of the Year
Since 1990, to recognize inventors of exceptional, cutting-edge NASA technologies, the NASA Inventions and Contributions Board has rewarded outstanding scientific and technical contributions through the NASA Invention of the Year Award.

NASA Software of the Year
Established in 1994, the NASA Software of the Year Award is given to those programmers and developers who have created outstanding software for the Agency.

Spinoff Year(s)	Technology	R&D 100	Space Technology Hall of Fame	NASA Invention of the Year	NASA Software of the Year
1976, 1977, 1981, 2002, 2005	Memory Foam		1998		
1976	Improved Firefighter's Breathing System		1988		
1977, 1979	Liquid-Cooled Garments		1993		
1978, 1990	Fabric Roof Structures		1989		
1978, 1981, 1983, 2005	Phase-Insensitive Ultrasonic Transducer	1978			
1981	Cordless Tools		1989		
1983, 1986	PMR-15 Polyimide Resin	1977	1991		

Spinoff Year(s)	Technology	R&D 100	Space Technology Hall of Fame	NASA Invention of the Year	NASA Software of the Year
1984, 1996	Scratch Resistant Lenses		1989		
1985, 2008	Redox Energy Storage System	1979			
1985	Safety Grooving		1990		
1985, 1986, 1987, 1991, 1993, 2003, 2008	Earth Resources Laboratory Applications Software (ELAS)		1992		
1987, 2008	VisiScreen (Ocular Screening System)	1987	2003		
1988	Sewage Treatment With Water Hyacinths		1988		
1989	Data Acquisition and Control System Model 9450/CAMAC	1986			
1990	NASA Structural Analysis (NASTRAN) Computer Software		1988		
1991	Heart Defibrillator Energy Source		1999		
1992	Dexterous Hand Master (DHM)	1989			
1994	Ballistic Electron Emission Microscope (BEEM)	1990			
1994, 2008	Omniview Motionless Camera	1993			
1994	Stereotactic Breast Biopsy Technology		1997		
1995	1100C Virtual Window	1994			
1995, 2006, 2008	Microbial Check Valve		2007		

Spinoff Year(s)	Technology	R&D 100	Space Technology Hall of Fame	NASA Invention of the Year	NASA Software of the Year
1996, 2004	Tetrahedral Unstructured Software System (TetrUSS)				1996, 2004
1996	Anti-Shock Trousers		1996		
1996	Automated Hydrogen Gas Leak Detector	1995			
1996	Ceramics Analysis and Reliability Evaluation of Structures/Life (CARES/Life)	1995			1994
1996	Memory Short Stack Semiconductor	1994			
1997	Foster-Miller Fiber Optic Polymer Reaction Monitor	1990			
1997, 2007	Power Factor Controller		1988		
1997	Reaction/Momentum Wheel: Apparatus for Providing Torque and for Storing Momentum Energy			1998	
1998	Data Matrix Symbology		2001		
1999	Active Pixel Sensor		1999		
1999	DeltaTherm 1000	1994			
1999	Superex Tube Extrusion Process	1995			
1999	Precision GPS Software System		1998, 2004		2000
1999	Process for Preparing Transparent Aromatic Polyimide Film			1999	
1999	Tempest Server	1999			1998

Spinoff Year(s)	Technology	R&D 100	Space Technology Hall of Fame	NASA Invention of the Year	NASA Software of the Year
2000	Genoa: A Progressive Failure Analysis Software System	2000			1999
2000	Humanitarian Demining Device		2003		
2001	Composite Matrix Resins and Adhesives (PETI-5)			1998	
2001, 2008	Flexible Aerogel Superinsulation	2003			
2001	LARC PETI-5 Polyimide Resin	1997		1998	
2001	The SeaWiFS Data Analysis System (SeaDAS)				2003
2001	TOR Polymers	2000			
2001	Video Image Stabilization and Registration (VISAR)		2001	2002	
2002	Automatic Implantable Cardiovertor Defibrillator		1991		
2002	DeBakey Rotary Blood Pump (Ventricular Assist Device-VAD)		1999	2001	
2002	Hybrid Ice Protection System	2003			
2002	Quantum Well Infrared Photodetector (QWIP)		2001		
2002	Virtual Window		2003		
2003	Cart3D				2002
2003	Cochlear Implant		2003		

Spinoff Year(s)	Technology	R&D 100	Space Technology Hall of Fame	NASA Invention of the Year	NASA Software of the Year
2003	Generalized Fluid System Simulation Program				2001
2003	LADARVision 4000		2004		
2003	MedStar Monitoring System		2004		
2003	Personal Cabin Pressure Altitude Monitor and Warning System			2003	
2004	InnerVue Diagnostic Scope System		2005		
2004	Land Information System (LIS) v. 4.0				2005
2004	NanoCeram Superfilters	2002	2005		
2004	Numerical Propulsion System Simulation (NPSS)				2001
2004	Outlast Smart Fabric Technology		2005		
2004	Photrodes for Electrophysiological Monitoring	2002			
2005	PS/PM300 High-Temperature Solid Lubricant Coatings	2003			
2005	iROBOT PackBOT Tactical Mobile Robot		2006		
2005, 2008	Light-Emitting Diodes for Medical Applications		2000		
2005	Numerical Evaluation of Stochastic Structures Under Stress (NESSUS)	2005			
2005	THUNDER Actuators	1996			

Spinoff Year(s)	Technology	R&D 100	Space Technology Hall of Fame	NASA Invention of the Year	NASA Software of the Year
2005	Zero-Valent Metal Emulsion for Reductive Dehalogenation of DNAPL-Phase Environmental Contaminants		2007	2005	
2006	Novariant RTK AutoFarm AutoSteer		2006		
2006	Radiant Barrier		1996		
2006	2006 Petroleum Remediation Product		2008		
2007	Advanced Lubricants		2000		
2007	Atomic Oxygen System for Art Restoration	2002			
2007	Future Air Traffic Management Concepts Evaluation Tool (FACET)				2006
2007	Macro-Fiber Composite Actuator	2000		2007	
2007	Portable-Hyperspectral Imaging Systems		2005		
2007	ResQPOD: Circulation-Enhancing Device		2008		
2007	ArterioVision: Noninvasive Cardiovascular Disease Detection		2008		
2008	DMBZ-15 High-Temperature Polyimide	2003			
2008	LaRC-SI: Soluble Imide	1995			
2008	Optical Backscatter Reflectometer	2007			

Executive Summary

In accordance with congressional mandates cited in the National Aeronautics and Space Act of 1958 and the Technology Utilization Act of 1962, NASA was directed to encourage greater use of the Agency's knowledge. For 50 years, NASA has nurtured partnerships with the private sector to facilitate the transfer of NASA-developed technologies. The benefits of these partnerships have reached throughout the economy and around the globe, as the resulting commercial products contributed to the development of services and technologies in the fields of health and medicine, transportation, public safety, consumer goods, environmental resources, computer technology, and industry.

Executive Summary

N ASA *Spinoff* highlights the Agency's most significant research and development activities and the successful transfer of NASA technology, showcasing the cutting-edge research being done by the Nation's top technologists and the practical benefits that come back down to Earth in the form of tangible products that make our lives better. The benefits featured in this year's issue include:

Health and Medicine

Robotics Offer Newfound Surgical Capabilities

Barrett Technology Inc., of Cambridge, Massachusetts, completed three Phase II Small Business Innovation Research (SBIR) contracts with Johnson Space Center, during which the company developed and commercialized three core technologies: a robotic arm, a hand that functions atop the arm, and a motor driver to operate the robotics. Among many industry uses, a recent adaptation of the arm has been cleared by the U.S. Food and Drug Administration (FDA) for use in a minimally invasive knee surgery procedure, where its precision control makes it ideal for inserting a very small implant.
page 46

In-Line Filtration Improves Hygiene and Reduces Expense

MRLB International Inc., of Fergus Falls, Minnesota, designed the DentaPure waterline purification cartridge using water purification research conducted by Umpqua Research Company, of Myrtle Creek, Oregon, as part of SBIR contracts from Johnson Space Center. Various models now address a variety of needs, and are used in dental offices and dental schools across the country. Currently the only waterline system recognized by the FDA as a medical device which meets all known standards and by the U.S. Environmental Protection Agency (EPA) as an antimicrobial device, DentaPure has also been utilized by the U.S. Air Force.
page 48

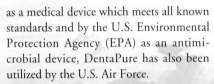

LED Device Illuminates New Path to Healing

Quantum Devices Inc., of Barneveld, Wisconsin, was granted a NASA SBIR contract to develop an LED light source for use in a surgical environment. Several SBIR contracts from Marshall Space Flight Center helped develop the High Emissivity Aluminiferous Light-emitting Substrate (HEALS), then successfully applied it in cases of pediatric brain tumors and the prevention of oral mucositis in pediatric bone marrow transplant patients. The HEALS and subsequent WARP 10 technologies have won many awards, including a "Tibbets Award." Recently, the next-generation device, the WARP 75, was released.
page 50

Polymer Coats Leads on Implantable Medical Device

Langley Research Center's Soluble Imide was licensed by Medtronic Inc., of Minneapolis, Minnesota, for use as insulation on thin metal wires connected to its implantable cardiac resynchronization therapy devices, for patients experiencing heart failure. The devices resynchronize the contractions of the heart's ventricles by sending tiny electrical impulses to the heart muscle, helping the heart pump blood throughout the body more efficiently.
page 52

Lockable Knee Brace Speeds Rehabilitation

Gary Horton, owner and operator of Horton's Orthotic Lab Inc., in Little Rock, Arkansas, was visiting Marshall Space Flight Center when he unexpectedly received assistance with a knee brace he was designing. Marshall engineers shared with him designs they had developed for a lockable joint with a hinge brake. Horton licensed the technology from Marshall and then set about applying the design concept to a new type of orthotic, a knee brace that automatically unlocks during the swinging phase of walking, but then is able to reengage for stability upon heel strike.
page 54

Robotic Joints Support Horses and Humans

Cable-compliant joints developed at Goddard Space Flight Center provided the key elements in the NASA Equine Support Technology (N.E.S.T.). The device supports a horse's weight with a special harness and controls the pelvis without restricting hip movement. Enduro Medical Technologies, of South Windsor, Connecticut, expects the N.E.S.T. to revolutionize treatment for horses with leg injuries. The human version of the technology, released in 2003, is currently assisting U.S. service personnel

rehabilitating from spinal cord or brain injuries at Walter Reed Army Medical Center, in Washington, DC.
page 56

Photorefraction Screens Millions for Vision Disorders

Marshall Space Flight Center scientists adapted optics technology for eye screening methods using a process called photorefraction, using a camera system with a specifically angled telephoto lens and flash to photograph a subject's eye. In 1991, NASA transferred the exclusive license for the system to Vision Research Corporation, of Birmingham, Alabama, and over 3 million children have been screened for vision disorders since.
page 58

Periodontal Probe Improves Exams, Alleviates Pain

Visual Programs Inc., of Richmond, Virginia, licensed the Periodontal Structures Mapping System from Langley Research Center. The resulting Ultrasonagraphic Probe (USProbe) is a noninvasive tool to make and record differential measurements of a patient's periodontal ligaments. The USProbe automatically detects, maps, and diagnoses problem areas by integrating diagnostic medical ultrasound techniques with advanced artificial intelligence. In addition to solving the problems associated with conventional probing, the USProbe may also provide information on the condition of the gum tissue and the quality and extent of the bond to the tooth surface.
page 60

Magnetic Separator Enhances Treatment Possibilities

Since 1988, NASA has issued over 25 SBIR contracts with 4 NASA centers to the company now known as Techshot Inc., of Greenville, Indiana. Currently, Techshot and a spinoff company, IKOTech, are marketing the Magsort, a Quadruple Magnetic Sorter, which collects specific biological cells from a liquid suspension by running it through a magnet assembly. Its applications include the detection of rare cancer cells in circulating blood and the removal of undesired cells from bone marrow transplants.
page 62

Transportation

Lithium Battery Power Delivers Electric Vehicles to Market

Hybrid Technologies Inc., a manufacturer and marketer of lithium-ion battery electric vehicles, based in Las Vegas, Nevada, and with research and manufacturing facilities in Mooresville, North Carolina, entered into a Space Act Agreement with Kennedy Space Center to determine the utility of lithium-powered fleet vehicles. NASA contributed engineering expertise for the car's advanced battery management system and tested a fleet of zero-emission vehicles on the Kennedy campus. Hybrid Technologies now offers a series of purpose-built lithium electric vehicles dubbed the LiV series, aimed at the urban and commuter markets.
page 66

Advanced Control System Increases Helicopter Safety

With support and funding from a Phase II NASA SBIR project from Ames Research Center, Hoh Aeronautics Inc. (HAI), of Lomita, California, produced HeliSAS, a low-cost, lightweight, attitude-command-attitude-hold stability augmentation system (SAS) for civil helicopters and unmanned aerial vehicles. HeliSAS proved itself in over 160 hours of flight testing and demonstrations in a Robinson R44 Raven helicopter, a commercial helicopter popular with news broadcasting and police operations. Chelton Flight Systems, of Boise, Idaho, negotiated with HAI to develop, market, and manufacture HeliSAS, now available as the Chelton HeliSAS Digital Helicopter Autopilot.
page 68

Aerodynamics Research Revolutionizes Truck Design

During the 1970s and 1980s, researchers at Dryden Flight Research Center conducted numerous tests to refine the shape of trucks to reduce aerodynamic drag and improve efficiency. During the 1980s and 1990s, a team based at Langley Research Center explored controlling drag and the flow of air around a moving body. Aeroserve Technologies Ltd., of Ottawa, Canada, with its subsidiary, Airtab LLC, in Loveland, Colorado, applied the research from Dryden and Langley to the development of the Airtab vortex generator. Airtabs create two counter-rotating vortices to reduce wind resistance and aerodynamic drag from trucks, trailers, recreational vehicles, and many other vehicles.
page 70

Engineering Models Ease and Speed Prototyping

System-response models developed by LMS International NV, a Belgium-based company with over 30 offices worldwide including Troy, Michigan, were used to calculate side-wall loads on J-2X nozzles at Marshall Space Flight Center. The J-2X will power the Ares launch vehicle—NASA's next-generation spacecraft. LMS engineers gained knowledge to be used for engineering applications in a wide range of other industries. By providing onsite support for tests, the LMS technical support and development staff seize opportunities like the work with NASA to expand their knowledge of tests and dynamics in real-world applications.

page 74

Software Performs Complex Design Analysis

Optimal Solutions Software LLC, of Provo, Utah, and Idaho Falls, Idaho, creates highly innovative engineering design improvement products to enable engineers to more reliably, creatively, and economically design new products in high-value markets. The company entered into an SBIR contract with Stennis Space Center, under which it extensively used its arbitrary shape deformation software to improve pressure loss, velocity, and flow quality in the pipes utilized by NASA. The product is available under the trade name Sculptor.

page 76

Public Safety

Space Suit Technologies Protect Deep-Sea Divers

Paragon Space Development Corporation is a Tucson, Arizona-based firm specializing in aerospace engineering and technology development, and is a major supplier of environmental control and life support system and subsystem design for the aerospace industry. Through its work with NASA, the company has developed a suit for protecting divers who are called on to work in extreme and dangerous conditions, such as high pressure, toxic chemical spills, the hot waters of the Persian Gulf, and chemical warfare agents.

page 80

Fiber Optic Sensing Monitors Strain and Reduces Costs

Luna Technologies, a division of Luna Innovations Incorporated, based in Blacksburg, Virginia, licensed technologies developed at Langley Research Center as part of the ultrasonic dynamic vector stress sensor. Luna released the Optical Vector Analyzer (OVA), Distributed Sensing System (DSS), and the Optical Backscatter Reflectometer (OBR) platforms. The OVA platform fiber optic sensing instruments include a set for linear characterization of single-mode optical components. The DSS and OBR platforms are two different techniques for distributed sensing: the DSS uses Fiber Bragg Gratings, and the OBR uses standard telecom-grade optical fiber.

page 82

Polymer Fabric Protects Firefighters, Military, and Civilians

In 1967, NASA contracted with Celanese Corporation, of New York, to develop a line of PBI textiles for use in space suits and vehicles. In 2005, the PBI fiber and polymer business was sold to PBI Performance Products Inc., of Charlotte, North Carolina, under the ownership of the InterTech Group, of North Charleston, South Carolina. PBI Performance Products now offers two distinct lines: PBI, the original heat and flame resistant fiber; and Celazole, a family of high-temperature PBI polymers available in true polymer form. PBI is now used in numerous firefighting, military, motor sports, and other applications.

page 84

Advanced X-Ray Sources Ensure Safe Environments

Ames Research Center awarded inXitu Inc. (formerly Microwave Power Technology), of Mountain View, California, an SBIR contract to develop a new design of electron optics for forming and focusing electron beams that is applicable to a broad class of vacuum electron devices. This technology offers an inherently rugged and more efficient X-ray source for material analysis; a compact and rugged X-ray source for smaller rovers on future Mars missions; and electron beam sources to reduce undesirable emissions from small, widely distributed pollution sources, and remediation of polluted sites.

page 86

Consumer, Home, and Recreation

Wireless Fluid-Level Measurement System Equips Boat Owners

While developing a measurement acquisition system to be used to retrofit aging aircraft with vehicle health monitoring capabilities, Langley Research Center developed an innovative wireless fluid-level measurement system. The NASA technology was of interest to Tidewater Sensors LLC, of Newport News, Virginia, because of its many advantages over conventional fuel management systems, including its ability to provide an accurate measurement of volume while a boat is experiencing any rocking motion due to waves or people moving about on the boat. These advantages led the company to license this novel fluid-level measurement system from NASA for marine applications.
page 90

Mars Cameras Make Panoramic Photography a Snap

The Mars rover Panoramic Mast Assemblies inspired scientists at Ames Research Center and Carnegie Mellon University to find more "down-to-Earth" photographic and virtual exploration applications for consumers. With the Austin, Texas-based Charmed Labs LLC, scientists created a prototype for the Gigapan robotic platform for consumer cameras, which automates the creation of highly detailed digital panoramas. The scientists also created a Web site and photographic stitching software to accompany the Gigapan platform.
page 92

Experiments Advance Gardening at Home and in Space

NASA research with BioServe Space Technologies and AgriHouse Inc., developing aeroponic gardening for space flight, inspired an innovative home gardening appliance. AeroGrow International Inc., of Boulder, Colorado, designed and released the AeroGarden line of countertop gardens based on NASA studies. One element, the Seed Pod, has since been used by BioServe as part of an experiment on the International Space Station, as its design would protect tomato seeds and prevent premature germination.
page 94

Space Age Swimsuit Reduces Drag, Breaks Records

Because of Langley Research Center's experience in studying the forces of friction and drag, Los Angeles-based SpeedoUSA asked the Agency to help design a swimsuit shortly after the 2004 Olympics. The LZR Racer reduces skin friction drag 24 percent more than the previous Speedo racing suit. The research seems to have paid off; in March 2008, athletes wearing the LZR Racer broke 13 world records.
page 96

Immersive Photography Renders 360° Views

An SBIR contract through Langley Research Center helped Interactive Pictures Corporation, of Knoxville, Tennessee, create an innovative imaging technology. This technology is a video imaging process that allows real-time control of live video data and can provide users with interactive, panoramic 360° views. The camera system can see in multiple directions, provide up to four simultaneous views, each with its own tilt, rotation, and magnification, yet it has no moving parts, is noiseless, and can respond faster than the human eye. In addition, it eliminates the distortion caused by a fisheye lens, and provides a clear, flat view of each perspective.
page 98

Historic Partnership Captures Our Imagination

Victor Hasselblad AB, of Gothenburg, Sweden, has enjoyed a long-lived collaboration with NASA, especially Johnson Space Center. For over four decades, Hasselblad has supplied camera equipment to the NASA Space Program, and Hasselblad cameras still take on average between 1,500 and 2,000 photographs on each space shuttle mission. Collaboration with NASA has allowed a very small company to achieve worldwide recognition—Hasselblad's operations now include centers in Parsippany, New Jersey; and Redmond, Washington; as well as France and Denmark—and consumer camera models have featured improvements resulting from refinements for the space models.
page 100

Outboard Motor Maximizes Power and Dependability

Developed by Jonathan Lee, a structural materials engineer at Marshall Space Flight Center, and PoShou Chen, a scientist with Huntsville, Alabama-based Morgan Research Corporation, MSFC-398 is a high-strength aluminum alloy able to operate at high temperatures. MSFC-398

was licensed for marine applications by Bombardier Recreational Products Inc., and is now found in the complete line of Evinrude E-TEC outboard motors, a line of two-stroke motors that maintain the power and dependability of a two-stroke with the refinement of a four-stroke.
page 104

Space Research Fortifies Nutrition Worldwide

NASA's Controlled Ecological Life Support Systems program attempted to address basic needs of crews, meet stringent payload and power usage restrictions, and minimize space occupancy, by developing living, regenerative ecosystems that would take care of themselves and their inhabitants. An experiment from this program evolved into one of the most widespread NASA spinoffs of all time—a method for manufacturing an algae-based food supplement that provides the nutrients previously only available in breast milk. Martek Biosciences Corporation, in Columbia, Maryland, now manufactures this supplement, and it can be found in over 90 percent of the infant formulas sold in the United States, as well as those sold in over 65 other countries. With such widespread use, the company estimates that over 24 million babies worldwide have consumed its nutritional additives.
page 106

Aerogels Insulate Missions and Consumer Products

Aspen Aerogels, of Northborough, Massachusetts, worked with NASA through an SBIR contract with Kennedy Space Center to develop a robust, flexible form of aerogel for cryogenic insulation for space shuttle launch applications. The company has since used the same manufacturing process developed under the SBIR award to expand its product offerings into the more commercial realms, making the naturally fragile aerogel available for the first time as a material that can be handled and installed just like standard insulation.
page 108

Environmental and Agricultural Resources

Computer Model Locates Environmental Hazards

Catherine Huybrechts Burton founded San Francisco-based Endpoint Environmental (2E) LLC in 2005 while she was a student intern and project manager at Ames Research Center with NASA's DEVELOP program. The 2E team created the Tire Identification from Reflectance Model, which algorithmically processes satellite images using turnkey technology to retain only the darkest parts of an image. This model allows 2E to locate piles of rubber tires, which often are stockpiled illegally and cause hazardous environmental conditions and fires.
page 112

Battery Technology Stores Clean Energy

Headquartered in Fremont, California, Deeya Energy Inc. is now bringing its flow batteries to commercial customers around the world after working with former Marshall Space Flight Center scientist, Lawrence Thaller. Deeya's liquid-cell batteries have higher power capability than Thaller's original design, are less expensive than lead-acid batteries, are a clean energy alternative, and are 10 to 20 times less expensive than nickel-metal hydride batteries, lithium-ion batteries, and fuel cell options.
page 114

Robots Explore the Farthest Reaches of Earth and Space

Deep Ocean Engineering (DOE) Inc., of San Leandro, California, received several SBIR awards from NASA to develop remotely operated vehicle (ROV) technologies with Ames Research Center. DOE engineers developed a concept for a versatile and robust locomotion methodology based on snake and worm morphologies. This "super snake" has the ability to transition seamlessly from one environment to another, such as land to water to burrowing into soft sediment. DOE ROVs are in use by U.S. armed forces, Hydro Quebec, and more than 40 universities and scientific organizations.
page 116

Portable Nanomesh Creates Safer Drinking Water

In 2003, Seldon Technologies Inc., of Windsor, Vermont, began designing a carbon Nanomesh for filtering impurities from drinking water. Testing in EPA-certified facilities showed that Seldon's filters removed more than 99 percent of bacteria and viruses, numerous chemical contaminants, and endotoxins, such as *Escherichia coli (E.coli)* and *Salmonella*. Using a carbon Nanomesh, the WaterStick filters about 5 gallons (200 milliliters) of water a minute simply using water pressure and gravity—without electricity, heat, chemical additives, or environmental impact.

page 118

Innovative Stemless Valve Eliminates Emissions

Big Horn Valve Inc. (BHVI), of Sheridan, Wyoming, won a series of SBIR and Small Business Technology Transfer (STTR) contracts with Kennedy Space Center and Marshall Space Flight Center to explore and develop a revolutionary valve technology. BHVI developed a low-mass, high-efficiency, leak-proof cryogenic valve using composites and exotic metals, and had no stem-actuator, few moving parts, with an overall cylindrical shape. The valve has been installed at a methane coal gas field, and future applications are expected to include in-flight refueling of military aircraft, high-volume gas delivery systems, petroleum refining, and in the nuclear industry.

page 120

Web-Based Mapping Puts the World at Your Fingertips

NASA's award-winning Earth Resources Laboratory Applications Software (ELAS) package was developed at Stennis Space Center. Since 1978, ELAS has been used worldwide for processing satellite and airborne sensor imagery data of the Earth's surface into readable and usable information. DATASTAR Inc., of Picayune, Mississippi, has used ELAS software in the DATASTAR Image Processing Exploitation (DIPEx) desktop and Internet image processing, analysis, and manipulation software. The new DIPEx Version III includes significant upgrades and improvements compared to its esteemed predecessor. A true World Wide Web application, this product evolved with worldwide geospatial dimensionality and numerous other improvements that seamlessly support the Web version.

page 122

Computer Technology

Program Assists Satellite Designers

Annapolis, Maryland-based designAmerica Inc., a small aerospace company specializing in the development and delivery of ground control systems for satellites and instrumentation, assisted Goddard Space Flight Center in the development of the ASIST software, a real-time command and control system for spacecraft development, integration, and operations. It was designed to be fully functional across a broad spectrum of satellites and instrumentation, while also being user friendly. The company now has rights to commercial use of the program and is offering it to government and industry satellite designers.

page 126

Water-Based Coating Simplifies Circuit Board Manufacturing

The Polymers Branch at Glenn Research Center's extensive knowledge of polyimide chemistry and its expertise in the synthesis of ultraviolet light curable polyimides were the critical components that allowed Advanced Coatings International, of Akron, Ohio, to prototype the platform chemistry for a polyimide-based, waterborne, liquid photoimagable coating ideal for the manufacture of printed circuit boards.

page 127

Software Schedules Missions, Aids Project Management

Through several long-term SBIR contracts, Knowledge-Based Systems Inc. (KBSI), of College Station, Texas, developed three advanced system management softwares: WorkSim, Model Mosaic, and AIOXFinder. Used independently or as a suite, these programs help manage complex projects and have been applied to several NASA missions.

page 128

Software Analyzes Complex Systems in Real Time

VIASPACE Inc., of Pasadena, California, licensed the Spacecraft Health Inference Engine (SHINE) software from NASA. It was designed to monitor, analyze, and diagnose real-time and non-real-time systems and, in addition to having been used on at least eight major NASA missions, has found application in the military and industrial realms.

page 130

Wireless Sensor Network Handles Image Data

Vexcel Corporation, of Boulder, Colorado, received STTR funding through Goddard Space Flight Center to develop wireless sensor network technology that now aids in the high-speed handling of image data. This technology has uses in both the commercial sector, where it is used to relay satellite imagery to the desktop, and in the government sector, where NASA

is finding continued use in terrestrial and interplanetary studies.
page 132

Virtual Reality System Offers a Wide Perspective

As part of an SBIR agreement to improve the telepresence of Johnson Space Center's Robonaut, Baltimore-based Sensics Inc. created a head-mounted display with a high-resolution, three-dimensional panorama. The Sensics piSight is now being sold commercially for high-end virtual reality applications. Virtual surroundings appear in the viewfinder and respond to head movements.
page 134

Software Simulates Sight: Flat Panel Mura Detection

Radiant Imaging Inc., of Duvall, Washington, licensed the Spatial Standard Observer (SSO) software from Ames Research Center. The SSO simulates a simplified model of human spatial vision, operating on a pair of images that are viewed at a specific viewing distance with pixels having a known relation to luminance. The SSO software was used to develop the TrueMURA Analysis Module, incorporated into Radiant Imaging's ProMetric 9.1 system. When used in conjunction with the ProMetric Series Imaging Colorimeters, the new software module provides a complete characterization and testing system for flat panel displays, especially LCD panels and displays.
page 136

Inductive System Monitors Tasks

The Inductive Monitoring System is software developed at Ames Research Center that uses artificial intelligence and data mining techniques to build system-monitoring knowledge bases from archived or simulated sensor data. This information is then used to detect unusual or anomalous behavior that may indicate an impending system failure. iSagacity Inc., based out of Portland, Maine, executed a nonexclusive license and is now offering it for use for water treatment plants, water heating and cooling in the process industry, oil refineries, public water distribution, and power generation plants.
page 138

Mars Mapping Technology Brings Main Street to Life

Berkeley, California-based earthmine inc., licensed 3-D data-generation software and algorithms from NASA's Jet Propulsion Laboratory originally used to create a 3-D representation of the local terrain to allow autonomous routing of the Mars Exploration Rovers. earthmine combined the software and algorithms with its unique capture hardware and Web delivery technology in a system that integrates the information to deliver accurate street-level geospatial data through a Web-based interface. Complete municipalities are collected through high-quality, 3-D panoramic images—including every road, alley, and freeway—to create a complete, consistent, and publicly accessible geospatial view of cities for official and commercial applications.
page 140

Intelligent Memory Module Overcomes Harsh Environments

3D Plus USA Inc., of McKinney, Texas, licensed Radiation Tolerant Intelligent Memory Stack technology from Langley Research Center for systems and methods to detect a failure event in field programmable gate arrays. In partnership with Langley, 3D Plus developed the first high-density and fast access time memory module tolerant of space radiation effects. This module decreases design complexity for space-based electronics requiring memory with its simple interface and internal radiation tolerance management. Expected applications include commercial or scientific geostationary missions and deep space scientific exploration, and high-reliability computing in other radiation-intensive environments like nuclear facilities.
page 142

Integrated Circuit Chip Improves Network Efficiency

Under a 2002 Space Act Agreement, Rockville, Maryland-based BAE Systems Inc. worked with Goddard Space Flight Center to create a SpaceWire-based application-specific integrated circuit (ASIC) chip for bridging existing space electronics and Goddard's new link-and-switch routers. BAE Systems' ASIC integrates easily into an onboard system and also decreases the part count, overall system complexity, ongoing costs, and power requirements for the system's board while also improving speed and reliability.
page 144

Industrial Productivity

Novel Process Revolutionizes Welding Industry

Glenn Research Center, Delphi Corporation, and the Michigan Research Institute entered into a project to study the use of Deformation Resistance Welding (DRW) in the construction and repair of stationary structures with multiple geometries and dissimilar materials, such as those NASA might use on the Moon or Mars. Traditional welding technologies are burdened by significant business and engineering challenges, including high costs of equipment and labor, heat-affected zones, limited automation, and inconsistent quality. DRW addresses each of those issues, while drastically reducing welding, manufacturing, and maintenance costs.
page 148

Sensors Increase Productivity in Harsh Environments

California's San Juan Capistrano-based Endevco Corporation licensed three patents for high-temperature, harsh-environment silicon carbide (Si-C) pressure sensors from Glenn Research Center. The company is exploring their use in government markets, as well as in commercial markets, including commercial jet testing, deep well-drilling applications where pressure and temperature increase with depth, and in automobile combustion chambers.
page 150

Portable Device Analyzes Rocks and Minerals

inXitu Inc., of Mountain View, California, entered into a Phase II SBIR contract with Ames Research Center to develop technologies for the next generation of scientific instruments for materials analysis. The work resulted in a sample handling system that could find a wide range of applications in research and industrial laboratories as a means to load powdered samples for analysis or process control. Potential industries include chemical, cement, inks, pharmaceutical, ceramics, and forensics. Additional applications include characterizing materials that cannot be ground to a fine size, such as explosives and research pharmaceuticals.
page 152

NASA Design Strengthens Welds

Friction Stir Welding (FSW) is a solid-state joining process—a combination of extruding and forging—ideal for use when the original metal characteristics must remain as unchanged as possible. While exploring ways to improve the use of FSW in manufacturing, engineers at Marshall Space Flight Center created technologies to address the method's shortcomings. MTS Systems Corporation, of Eden Prairie, Minnesota, discovered the NASA-developed technology and then signed a co-exclusive license agreement to commercialize Marshall's design for use in high-strength structural alloys. The resulting process offers the added bonuses of being cost-competitive, efficient, and most importantly, versatile.
page 154

Polyimide Boosts High-Temperature Performance

Maverick Corporation, of Blue Ash, Ohio, licensed DMBZ-15 polyimide technology from Glenn Research Center. This ultrahigh-temperature material provides substantial weight savings and reduced machining costs compared to the same component made with more traditional metallic materials. DMBZ-15 has a wide range of applications from aerospace (aircraft engine and airframe components, space transportation systems, and missiles) to non-aerospace (oil drilling, rolling mill), and is particularly well-suited to use as face sheets with honey cones or thermal protection systems for reusable launch vehicles, which encounter elevated temperatures during launch and reentry.
page 156

NASA Innovation Builds Better Nanotubes

Nanotailor Inc., based in Austin, Texas, licensed Goddard Space Flight Center's unique single-walled carbon nanotube (SWCNT) fabrication process with plans to make high-quality, low-cost SWCNTs available commercially. Carbon nanotubes are being used in a wide variety of applications, and NASA's improved production method will increase their applicability in medicine, microelectronics, advanced materials, and molecular containment. Nanotailor built and tested a prototype based on Goddard's process, and is using this technique to lower the cost and improve the integrity of nanotubes, offering a better product for use in biomaterials, advanced materials, space exploration, highway and building construction, and many other applications.
page 158

NASA Technologies Enhance Our Lives

International Space Station

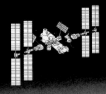

Space Telescopes and Deep Space Exploration

Satellites and Imaging Technology

Innovative technologies from NASA's space and aeronautics missions (above) transfer as benefits to many sectors of society (below).

Each benefit featured in *Spinoff* 2008 is listed with an icon that corresponds to the mission from which the technology originated. These NASA-derived technologies, when transferred to the public sector:

Health and Medicine

 Offer newfound surgical capabilities
page 46

 Improve hygiene and reduce expense
page 48

 Illuminate new paths to healing
page 50

 Coat leads on implantable medical devices
page 52

 Speed rehabilitation
page 54

 Support horses and humans
page 56

 Screen millions for vision disorders
page 58

 Improve exams, alleviate pain
page 60

 Enhance treatment possibilities
page 62

Transportation

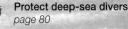 Deliver electric vehicles to market
page 66

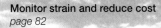 Increase helicopter safety
page 68

 Revolutionize truck design
page 70

 Ease and speed prototyping
page 74

 Perform complex design analysis
page 76

Public Safety

 Protect deep-sea divers
page 80

 Monitor strain and reduce cost
page 82

Protect firefighters, military, and civilians
page 84

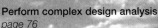 Ensure safe environments
page 86

Consumer, Home, and Recreation

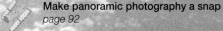 Equip boat owners
page 90

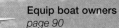 Make panoramic photography a snap
page 92

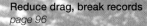 Advance gardening at home and in space
page 94

 Reduce drag, break records
page 96

 Render 360° views
page 98

 Capture our imagination
page 100

 Maximize power and dependability
page 104

 Fortify nutrition worldwide
page 106

 Insulate missions and consumer products
page 108

Space Transportation

Astronaut Life Support

Aeronautics Research

Environmental and Agricultural Resources

 Locate environmental hazards
page 112

 Store clean energy
page 114

 Explore the farthest reaches of Earth and space
page 116

 Create safer drinking water
page 118

 Eliminate emissions
page 120

 Put the world at your fingertips
page 122

Computer Technology

 Assist satellite designers
page 126

 Simplify circuit board manufacturing
page 127

 Schedule missions, aid project management
page 128

 Analyze complex systems in real time
page 130

 Handle image data
page 132

 Offer a wide perspective
page 134

Simulate sight
page 136

 Monitor tasks
page 138

Bring Main Street to life
page 140

Overcome harsh environments
page 142

Improve network efficiency
page 144

Industrial Productivity

 Revolutionize the welding industry
page 148

 Increase productivity in harsh environments
page 150

 Analyze rocks and minerals
page 152

 Strengthen welds
page 154

 Boost high-temperature performance
page 156

 Build better nanotubes
page 158

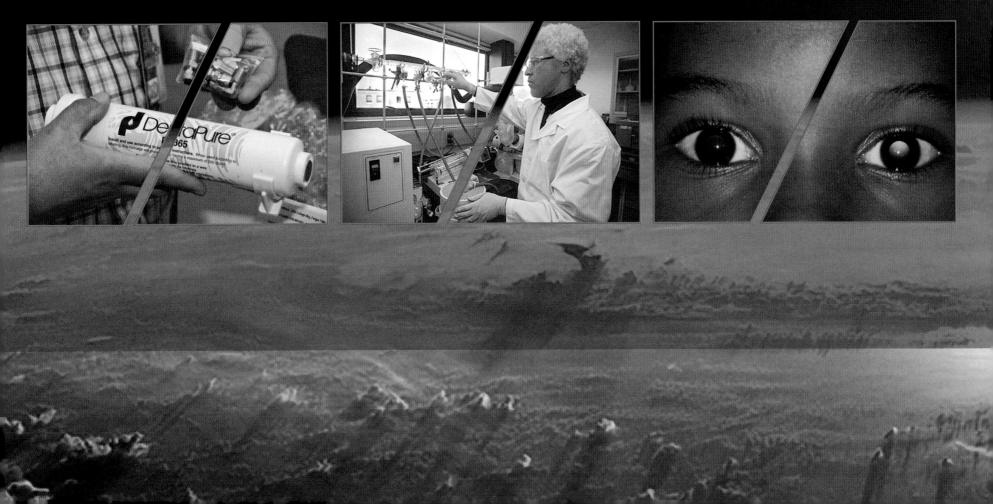

NASA Technologies Benefiting Society

The National Aeronautics and Space Act of 1958 required that NASA disseminate its information to the public, and the Technology Utilization Act of 1962 formalized the process through which the Agency was to accomplish this task. Today, NASA continues to seek industry partnerships to develop technologies that apply to NASA mission needs, provide direct societal benefits, and contribute to competitiveness in global markets. As part of NASA's mission, the Agency facilitates the transfer and commercialization of NASA-sponsored research and technology. These efforts not only support NASA, they enhance the quality of life in our hospitals, homes, and communities.

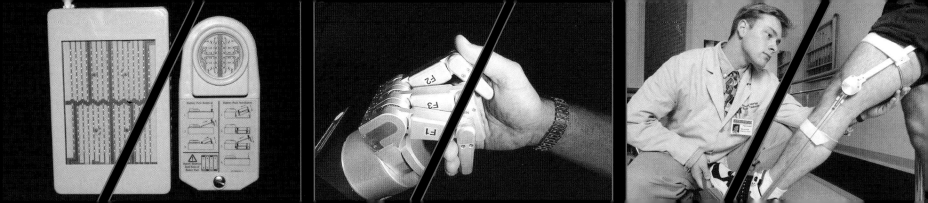

NASA research drives innovation that improves health care.
The technologies featured in this section:

- Offer newfound surgical capabilities
- Improve hygiene and reduce expense
- Illuminate new paths to healing
- Coat leads on implantable medical devices
- Speed rehabilitation
- Support horses and humans
- Screen millions for vision disorders
- Improve exams, alleviate pain
- Enhance treatment possibilities

Robotics Offer Newfound Surgical Capabilities

Originating Technology/NASA Contribution

Anyone who has ever worked on a car's engine or tried to fix a sink knows the frustration of trying to perform precision work in a hard-to-reach place. Imagine how that sense of frustration might magnify when, instead of trying to wrap the head of a wrench around a leaky nut under the kitchen counter, the scenario involves conducting repairs on the International Space Station while floating nearly 200 miles above Earth. To ease this frustration, NASA funded work on autonomous robotic devices that would be able to retrieve tools and even crew outside of the station.

Partnership

Barrett Technology Inc., of Cambridge, Massachusetts, completed three Phase II **Small Business Innovation Research (SBIR)** contracts with Johnson Space Center. In 1989, the company worked with NASA on a Phase II to create a robotic arm, and in 1991, was again awarded a Phase II to create a hand. Nearly a decade later, the company was awarded a third Phase II for further miniaturization of the components that comprise the robotic devices.

Product Outcome

Barrett has developed and commercialized three core technologies that trace their roots directly back to the SBIR work with NASA. The first is a robotic arm, the whole-arm manipulation (WAM) system; the second is a hand that functions atop the arm, the BH8-Series; and the third is a motor driver that the company refers to as "the puck," as it is similar in shape to a hockey puck, but one-tenth the size.

The SBIR work with NASA led to the development of the first commercially available cable-driven robot, a distinction that earned Barrett a place in the "Guinness World Records" book as the world's most advanced robotic arm. Designed for applications that require superior adaptability, programmability, and dexterity, the WAM can reach around large objects and grasp them with its arm links like huge fingers, while conventional robotic arms are restricted to hand end-effectors, and thus restricted to grasping smaller objects. The WAM has other advantages over traditional robotic arms, in that it uses gear-free cables to manipulate its joints, allowing it to feel and control subtle forces.

The arm consists of a shoulder that operates on a gearless differential mechanism, an upper arm, and a gear-free elbow, forearm, and wrist. This arrangement of joints coincides with the human shoulder and elbow, but with much greater range of motion. Like a person's arm, but unlike any industrial robotic arm, the WAM Arm is backdriveable, meaning that any contact force along the arm or its hand is immediately felt at the motors, supporting graceful control of interactions with walls, objects, and even people. With a human-scale 3-foot reach, it is so quick that it can grab a major-league fastball, yet so sensitive that it responds to the gentlest touch. The WAM Arm is available in two main configurations, four-degrees-of-freedom and seven-degrees-of-freedom, both with human-like kinematics. Internally protected channels allow the user to pass electric lines and fiber optics required for custom end-effectors and sensors.

These characteristics make it ideal for myriad applications, including in space, where use of robots is often safer than people, in manufacturing, and in medicine.

Recently, an adaptation of the WAM has been cleared by the U.S. Food and Drug Administration for use in a minimally invasive knee surgery procedure, where its precision control makes it ideal for inserting a very small implant. Barrett Technology licensed the arm to MAKO Surgical Corporation, of Fort Lauderdale, Florida, for use in the company's "keyhole" orthopedic surgery procedures.

The company uses small titanium knee implants instead of the more common and more traumatic total-knee replacement, thus limiting surgery to only the

The BarrettHand is a multifingered programmable grasper with the dexterity to secure target objects of different sizes, shapes, and orientations.

diseased part of the bone. These surgeries, however, require an array of complex implant shapes to cover the variety of disease patterns and bone geometries. To insert these devices, surgeons generally cut small pockets into the diseased bone with a high-speed, hand-held cutting device. While lacking in robotic precision, this technique provides the surgeon with the tactile sensation of the cutting, which in turn provides a wealth of intuitive information about the diseased bone that is not available from preoperative X-rays.

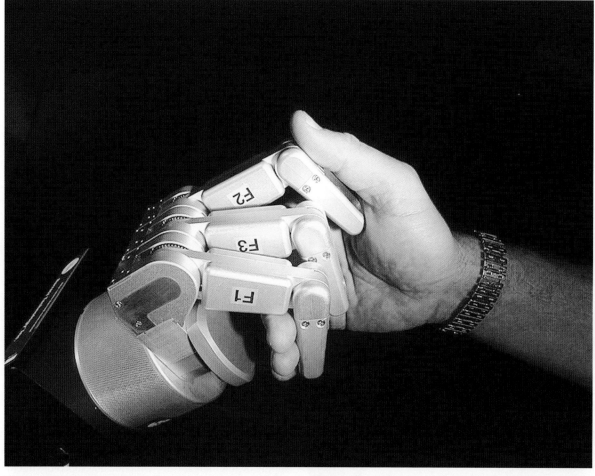

The BarrettHand offers unmatched versatility, programmability, and ease of integration with commercial robotic arms, including the Barrett WAM Arm.

and compact form, it is totally self-contained. Plus, communicating by industry-standard serial communications, integration with any robotic arm is fast and simple.

The BarrettHand BH8-Series neatly houses a CPU, software, communications electronics, servo-controllers, and four brushless motors. Of its three multijointed fingers, two have an extra degree of freedom with 180 degrees of synchronous lateral mobility supporting a large variety of grasp types.

Barrett's Ultra-Miniature Puck Brushless Servo Electronics Module, or "Puck," is the world's smallest and most power-efficient, high-performance servomotor controller. It is based on the work done with NASA, as well as grants from the U.S. Department of Energy and the National Science Foundation (NSF). Barrett has been shipping the Pucks in all of its robotic WAM Arms for the past 3 years, because the device offers several distinct advantages: the absence of a controller cabinet improves reliability and portability; the incredibly low power consumption (an order of magnitude less than any other arm in its class) increases safety and portability, while also making the device "greener;" and the ultra-high brushless-servo performance enables applications such as force-field-enabled medical surgery.

While the Puck is not currently available outside the WAM Arm today, Barrett has applied for NSF funding to develop features to make this module universally adaptable within 3 years to a wide range of brushless-servomotor applications. ❖

WAM™, BarrettHand™, and Puck™ are trademarks of Barrett Technology Inc.

The WAM-based technologies combine the best of a surgeon's intuition and a robot's precision through active haptics (touch sensing). The WAM Arm improves the precision of the implant pockets while still allowing the surgeon to feel bone condition. Matching pocket and implant geometries minimizes trauma, ensures secure implant retention, and optimizes resulting joint functionality.

Like the WAM Arm, the BH8-262 BarrettHand offers many benefits in dexterity. A multifingered programmable grasper, the BarrettHand can pick up objects of different sizes, shapes, and orientations. According to the company, integrating this device immediately multiplies the value of any arm requiring flexible automation. Even with its low weight (1.18kg)

In-Line Filtration Improves Hygiene and Reduces Expense

Originating Technology/NASA Contribution

Water, essential to sustaining life on Earth, is that much more highly prized in the unforgiving realm of space travel and habitation. Given a launch cost of $10,000 per pound for space shuttle cargo, however, each gallon of water at 8.33 pounds quickly makes Chanel No. 5 a bargain at $25,000 per gallon. Likewise, ample water reserves for drinking, food preparation, and bathing would take up an inordinate amount of storage space and infrastructure, which is always at a premium on a vessel or station.

Water rationing and recycling are thus an essential part of daily life and operations on the space shuttles and International Space Station (ISS). In orbit, where Earth's natural life support system is missing, the ISS itself has to provide abundant power, clean water, and breathable air at the right temperature and humidity for the duration of human habitation and with virtually no waste. The Environmental Control and Life Support System (ECLSS), under continuing developmet at the Marshall Space Flight Center, helps astronauts use and reuse their precious supplies of water. Future work will explore air management, thermal control, and fire suppression—in short, all of the things that will make human habitation in space comfortable and safe.

The ECLSS Water Recycling System (WRS), developed at Marshall, reclaims wastewaters from humans and lab animals in the form of breath condensate, urine, hygiene and washing, and other wastewater streams. On Earth, biological wastewater is physically filtered by granular soil and purified as microbes in the soil break down urea, converting it to a form that plants can absorb and use to build new tissue. Wastewater also evaporates and returns as fresh rain water—a natural form of distillation. WRS water purification machines on the ISS mimic these processes, though without microbes or the scale of these processes.

Partnership

Umpqua Research Company, of Myrtle Creek, Oregon, supplier of the bacterial filters used in the life support backpacks worn by space-walking astronauts, received a number of **Small Business Innovation Research (SBIR)** contracts from the Johnson Space Center to develop air and water purification technologies for human missions in space. A natural choice for water purification research, Umpqua has also provided the only space-certified and approved-for-flight water purification system, which has flown on all shuttle missions since 1990.

To prevent back-contamination of a drinking water supply by microorganisms, Umpqua developed the microbial check valve, consisting of a flow-through cartridge containing iodinated ion exchange resin. In addition to the microbial contact kill, the resin was found to impart a biocidal residual elemental iodine concentration to the water. Umpqua's valve and resin system was adopted by NASA as the preferred means of disinfecting drinking

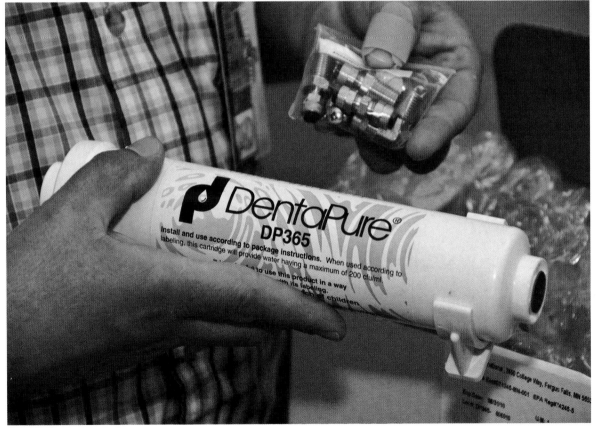

The DentaPure waterline purification cartridge sees use in 40 percent of dental schools in the United States and is lauded for offering remarkable filtration and significant cost savings.

water aboard U.S. spacecraft, and canisters are now used on space shuttle missions, the ISS, and for ground-based testing of closed life support technology. Iodine was selected by NASA as the disinfectant of choice because of its lower vapor pressure and reduced propensity for formation of disinfection byproducts compared to chlorine or bromine.

Product Outcome

MRLB International Inc., of Fergus Falls, Minnesota, used Umpqua's water purification technology in the design of the DentaPure waterline purification cartridge (*Spinoff* 1998). The cartridge incorporated a resin technology developed by private sector commercialization of Umpqua's system developed under NASA contract. NASA "was an excellent resource," stated Barry Hammarback, president and CEO of MRLB, "and greatly assisted our transition of the iodinated resin technology to the dental industry." DentaPure was designed to clean and decontaminate water as a link between filter and high-speed dental tools and other instruments, and offers easy installation on all modern dental unit waterlines with weekly replacement cycles. The product, like its NASA forebear, furnished disinfected water and maintained water purity even with "suckback," an effect caused by imperfect anti-retraction valves in dental instruments, which draws blood, saliva, and other materials from a patient's mouth into the waterline.

Since its appearance in *Spinoff* 1998, MRLB has continued to use the research conducted by Umpqua to further develop and refine its DentaPure in-line filters. Various models now address a variety of needs, and are used in dental offices and schools across the country. MRLB has paid particular attention to extending the life in lower water usage units—products that before touted a service/replacement interval of 7 days now require changing once every 40 to 365 days. In addition, DentaPure offers remarkable filtration: registered to provide 200 CFU/ml purity (Colony

Long service intervals of up to a year and ease of installation make the DentaPure cartridge particularly user-friendly.

Forming Unit/milliliter, a standard measure of microbial concentration)—the Centers for Disease Control and Prevention (CDC) standard is 500 CFU, and untreated lines can harbor in excess of 1,000,000 CFU/ml.

Continued evolution and improvement has led to many unique certifications and commendations for DentaPure. Currently, the only waterline system recognized by the U.S. Food and Drug Administration (FDA) as a medical device which meets all known standards, and by the U.S. Environmental Protection Agency (EPA) as an antimicrobial device, DentaPure has also been tested and utilized by the U.S. Air Force and dental schools in the United States and Europe, and was recognized by Clinical Research Associates as "Outstanding Product 2005."

Better filtration, greater capacity, and longer service intervals have also led to great savings—the University of Maryland Dental School estimates it saves $274,000 per year courtesy of DentaPure. The DentaPure system has proven so effective that 40 percent of dental schools nationwide employ it. Dr. Louis DePaola of the University of Maryland affirms, "The biggest benefit is that we have a system that is efficacious and user-friendly, in that it allows us to consistently deliver water that meets or exceeds CDC standards with a minimum of staff interaction—attach the unit and except for periodic monitoring you don't have to do anything for a year. It's very cost-effective—for a large institution like ours with an excess of 300 units, a daily or weekly treatment is not practical."

DentaPure is currently the number one product for constant chemical treatment applications in dentistry, and was the first treatment to meet CFU standards without interim cleaning protocols—the primary means by which it saves money. Turning to the future, Hammarback sees DentaPure "looking at remote site water purification for continuous use, providing yet longer lasting devices, and increasing product recycling." Ten years after *Spinoff* first profiled the many benefits of this technology it is utilized every day in myriad dental offices, schools, and labs, saving hundreds of thousands of dollars a year for users such as the University of Maryland. The investment in water filtration for space missions continues to pay huge dividends to users and society, year after year, in technologies so woven into our lives that we use them without even thinking about them. ❖

DentaPure® is a registered trademark of MRLB International Inc.

LED Device Illuminates New Path to Healing

Originating Technology/NASA Contribution

Among NASA's research goals is increased understanding of factors affecting plant growth, including the effects of microgravity. Impeding such studies, traditional light sources used to grow plants on Earth are difficult to adapt to space flight, as they require considerable amounts of power and produce relatively large amounts of heat. As such, an optimized experimental system requires much less energy and reduces temperature variance without negatively affecting plant growth results.

Ronald W. Ignatius, founder and chairman of the board at Quantum Devices Inc. (QDI), of Barneveld, Wisconsin, proposed using light-emitting diodes (LEDs) as the photon source for plant growth experiments in space. This proposition was made at a meeting held by the Wisconsin Center for Space Automation and Robotics, a NASA-sponsored research center that facilitates the commercialization of robotics, automation, and other advanced technologies. The Wisconsin group teamed with QDI to determine whether an LED system could provide the necessary wavelengths and intensities for photosynthesis, and the resultant system proved successful. The center then produced the Astroculture3, a plant growth chamber that successfully incorporated this LED light source, which has now flown on several space shuttle missions.

NASA subsequently identified another need that could be addressed with the use of LEDs: astronaut health. A central concern in astronaut health is maintaining healthy growth of cells, including preventing bone and muscle loss and boosting the body's ability to heal wounds—all adversely affected by prolonged weightlessness. Thus, having determined that LEDs can be used to grow plants in space, NASA decided to investigate whether LEDs might be used for photobiomodulation therapy (PBMT).

PBMT is an emerging medical and veterinary technique in which exposure to high-intensity, wavelength-specific light can stimulate or inhibit cellular function. PBMT modulates a body's organelles—structures within a cell (e.g., mitochondria, vacuoles, and chloroplasts) that store food, discharge waste, produce energy, or perform other functions analogous to the role of organs in the body

Studies have shown red LEDs are a viable light source for growing plants in space flight due to their small mass and volume, wavelength specificity, longevity, and safe operation.

as a whole—with wavelength-specific photon energy to increase respiratory metabolism, reduce the natural inflammatory response, accelerate recovery of injury or stress at the cellular level, and increase circulation.

Partnership

A NASA **Small Business Innovation Research (SBIR)** contract was granted to QDI to develop an LED light source for use in a surgical environment as the photon source for its proprietary Photodynamic Therapy (PDT) treatment. An emerging cancer treatment, PDT requires high-intensity, monochromatic light to turn on the cancer-killing properties of a drug, allowing physicians to activate a drug in the tumor only.

QDI and Dr. Harry T. Whelan of the Medical College of Wisconsin (known for groundbreaking research in PBMT) based their work on QDI's High Emissivity Aluminiferous Light-emitting Substrate (HEALS) technology, which was developed for use in the plant growth experiments in 1993. Several SBIR contracts from NASA's Marshall Space Flight Center between 1995 and 1998 helped QDI continue the evolution of HEALS in collaboration with Whelan, and the technology was successfully applied in cases of pediatric brain tumors and the prevention of oral mucositis in pediatric bone marrow transplant patients.

QDI then used a Defense Advanced Research Projects Agency (DARPA) SBIR contract to develop the WARP 10 (Warfighter Accelerated Recovery by Photobiomodulation) unit as a full realization of its PBMT research. WARP 10, a hand-held, portable, HEALS technology originally intended for military first aid applications, received U.S. Food and Drug Administration (FDA) clearance in 2003, and a consumer version was introduced for temporary relief of minor muscle and joint pain. WARP 10 has been found to relieve arthritis, muscle spasms, and stiffness; promote relaxation of tissue; and temporarily increase local blood circulation.

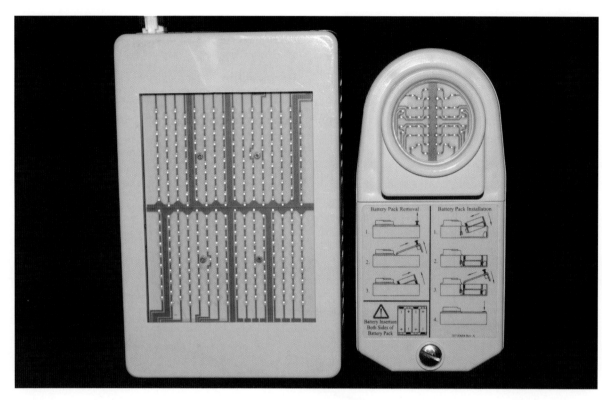

The WARP 75 (left) has 7.5 times more coverage area than the previously developed WARP 10 (right). The FDA has cleared the WARP 75 for the treatment of chronic pain, including the temporary relief of minor muscle and joint pain, arthritis, muscle spasms, and stiffness, by promoting relaxation of muscle tissue and temporarily increasing local blood circulation.

Product Outcome

Since HEALS and WARP 10 were originally profiled in *Spinoff* 2005, a flurry of activity has seen this unique technology showered in awards and the next-generation LED device gain FDA clearance and enter the market.

The HEALS and WARP 10 technologies have accrued an impressive résumé; the list of accolades received includes induction into the Space Technology Hall of Fame in 2000; being named a Marshall Space Flight Center "Hallmark of Success" as an outstanding commercialization of an SBIR-developed technology in 2004; and winning first place in the Wisconsin "Governor's New Product Awards" in 2005 for the development of WARP 10. The greatest and most recent accolade came in 2006, when QDI was nominated for and received a "Tibbetts Award."

Named for Roland Tibbetts, the acknowledged "father of the SBIR program," it is an annual government-wide award for small firms, projects, organizations, and individuals judged to exemplify the very best in SBIR achievement. These prestigious national awards emphasize recognizing those accomplishments where, in the judgment of those closely involved and often most immediately affected, the stimulus of SBIR funding has made an important and definable difference. Economic impact of technological innovation, business achievement, effective collaborations, demonstrated state and regional impact, and proven support are the main considerations.

On the heels of this honor, 2007 saw FDA clearance for the new WARP 75 device, the latest iteration of the technology that began with the HEALS technology. The WARP 75 improves on the WARP 10 design, boasting 7.5 times the actual coverage area of the WARP 10 ($75cm^2$ versus $10cm^2$), an automatic timed cycle of 88 seconds with an audible alarm, AC power, the ability to be mounted on an articulated arm, and fan cooling. System controls are located on the top panel for easy light dose delivery, and the device is placed directly against the skin where treatment is desired. The unit can be operated with one hand and remains cool to the touch during operation.

The WARP 75 continues the legacy of its predecessors in clinical trials with the Medical College of Wisconsin and the University of Alabama at Birmingham for the amelioration of oral mucositis pain in bone marrow transplant patients. QDI is exploring other medical applications of the HEALS-based technology, including combating the symptoms of bone atrophy, multiple sclerosis, diabetic complications, Parkinson's disease, and a variety of ocular applications. Most recently, Marshall awarded QDI another grant to study synergistic wound healing and conduct a PDT study with silver nanoclusters. Through all of their work, QDI remains dedicated to the principle that light provides the power for all life on Earth, and the belief that the quality, delivery, and control of light is essential to the wellness of the human race and our advancement into the future. ❖

HEALS®, WARP 10®, and WARP 75® are registered trademarks, and Photodynamic Therapy™ is a trademark of Quantum Devices Inc.

Polymer Coats Leads on Implantable Medical Device

Originating Technology/NASA Contribution

Langley Research Center's Soluble Imide (LaRC-SI) was discovered by accident. While researching resins and adhesives for advanced composites for high-speed aircraft, Robert Bryant, a Langley engineer, noticed that one of the polymers he was working with did not behave as predicted. After putting the compound through a two-stage controlled chemical reaction, expecting it to precipitate as a powder after the second stage, he was surprised to see that the compound remained soluble. This novel characteristic ended up making this polymer a very significant finding, eventually leading Bryant and his team to win several NASA technology awards, and an "R&D 100" award.

The unique feature of this compound is the way that it lends itself to easy processing. Most polyimides (members of a group of remarkably strong and incredibly heat- and chemical-resistant polymers) require complex curing cycles before they are usable. LaRC-SI remains soluble in its final form, so no further chemical processing is required to produce final materials, like thin films and varnishes. Since producing LaRC-SI does not require complex manufacturing techniques, it has been processed into useful forms for a variety of applications, including mechanical parts, magnetic components, ceramics, adhesives, composites, flexible circuits, multilayer printed circuits, and coatings on fiber optics, wires, and metals.

Bryant's team was, at the time, heavily involved with the aircraft polymer project and could not afford to further develop the polymer resin. Believing it was worth further exploration, though, he developed a plan for funding development and submitted it to Langley's chief scientist, who endorsed the experimentation. Bryant then left the high-speed civil transport project to develop LaRC-SI. The result is an extremely tough, lightweight thermoplastic that is not only solvent-resistant, but also has the ability to withstand temperature ranges from cryogenic levels to above 200 °C. The thermoplastic's unique characteristics lend it to many commercial applications; uses that Bryant believed would ultimately benefit industry and the Nation. "LaRC-SI," he explains, "is a product created in a government laboratory, funded with money from the tax-paying public. What we discovered helps further the economic competitiveness of the United States, and it was our goal to initiate the technology transfer process to ensure that our work benefited the widest range of people."

Several NASA centers, including Langley, have explored methods for using LaRC-SI in a number of applications from radiation shielding and as an adhesive to uses involving replacement of conventional rigid circuit boards. In the commercial realm, LaRC-SI can now be found in several commercial products, including the thin-layer composite unimorph ferroelectric driver and sensor (THUNDER) piezoelectric actuator, another "R&D 100" award winner (*Spinoff* 2005).

Partnership

Working with the Innovative Partnerships Program office at Langley, Medtronic Inc., of Minneapolis, Minnesota, licensed the material. This material has been evaluated for space applications, high-performance composites, and harsh environments; however, this partnership represents the first time that the material has been used in a medical device.

According to Bryant, "This partnership validates the belief we had that LaRC-SI needed to be introduced in (or by) the private sector: Lives can be saved and enhanced

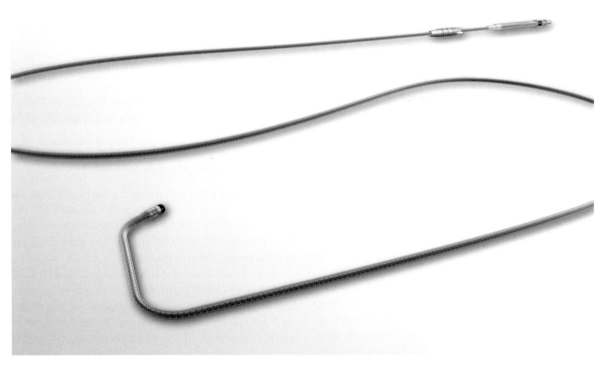

Medtronic's cardiac resynchronization therapy devices use the NASA-developed polymer as insulation on thin metal lead wires.

because we were able to develop our laboratory findings and provide public access to the material."

Product Outcome

Medtronic is the world leader in medical technology providing lifelong solutions for people with chronic disease. It offers products, therapies, and services that enhance or extend the lives of millions of people. Each year, 6 million patients benefit from Medtronic's technology, used to treat conditions such as diabetes, heart disease, neurological disorders, and vascular illnesses.

The company is testing the material for use as insulation on thin metal wires connected to its implantable cardiac resynchronization therapy (CRT) devices for patients experiencing heart failure, which resynchronize the contractions of the heart's ventricles by sending tiny electrical impulses to the heart muscle, helping the heart pump blood throughout the body more efficiently.

"Our work with NASA Langley was very collaborative," said Lonny Stormo, Medtronic vice president of therapy delivery research and development. "Our scientists discussed Medtronic's material requirements and NASA shared what it knows about the compound's properties as we continued our testing and evaluations."

In March 2007, Medtronic conducted the first clinical implants in the United States and Canada of the Medtronic over-the-wire lead (Model 4196), a dual-electrode left ventricular (LV) lead for use in heart failure patients with cardiac resynchronization therapy devices.

"Through this partnership, Medtronic was able to deliver a product with enhanced material properties," said Stormo. "In turn this helps our patients, which is the core of Medtronic's mission."

Placing a lead in the LV is widely recognized by physicians as the most challenging aspect of implanting CRT devices. Anatomic challenges can make it difficult to access and work within the coronary sinus to place a lead in the desired vein of the LV. The lead is specially designed for optimal tracking over a guide wire, which is

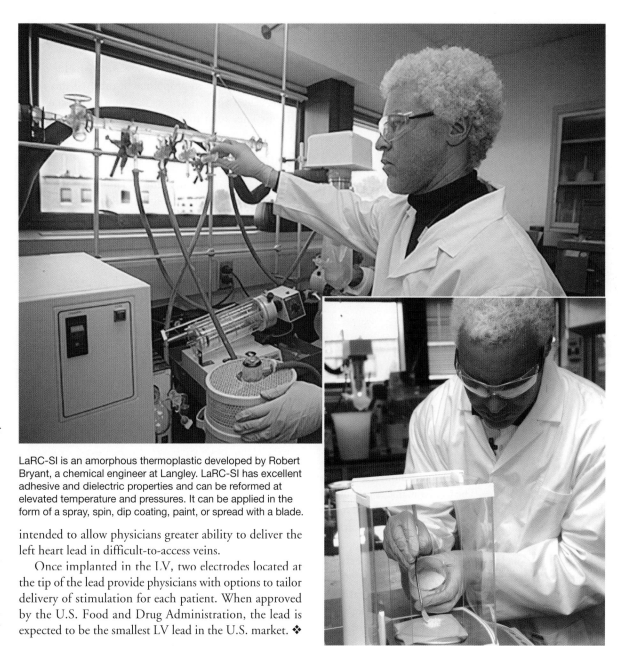

LaRC-SI is an amorphous thermoplastic developed by Robert Bryant, a chemical engineer at Langley. LaRC-SI has excellent adhesive and dielectric properties and can be reformed at elevated temperature and pressures. It can be applied in the form of a spray, spin, dip coating, paint, or spread with a blade.

intended to allow physicians greater ability to deliver the left heart lead in difficult-to-access veins.

Once implanted in the LV, two electrodes located at the tip of the lead provide physicians with options to tailor delivery of stimulation for each patient. When approved by the U.S. Food and Drug Administration, the lead is expected to be the smallest LV lead in the U.S. market. ❖

Lockable Knee Brace Speeds Rehabilitation

Originating Technology/NASA Contribution

Marshall Space Flight Center develops key transportation and propulsion technologies for the Space Agency. The Center manages propulsion hardware and technologies of the space shuttle, develops the next generation of space transportation and propulsion systems, oversees science and hardware development for the International Space Station, manages projects and studies that will help pave the way back to the Moon, and handles a variety of associated scientific endeavors to benefit space exploration and improve life here on Earth.

It is a large and diversified center, and home to a great wealth of design skill. Some of the same mechanical

The selectively lockable knee brace allows the knee to function while supporting the leg.

design skill that made its way into the plans for rocket engines and advanced propulsion at this Alabama-based NASA center also worked its way into the design of an orthotic knee joint that is changing the lives of people with weakened quadriceps.

Partnership

Gary Horton, owner and operator of Horton's Orthotic Lab Inc., in Little Rock, Arkansas, was visiting Marshall on unrelated business, when he unexpectedly received assistance with a knee brace he was designing.

He was attending a meeting at the Center, where, once the engineers learned he was an orthotist, they shared with him plans for several newly designed knee joints.

The particular design that caught his eye was one by Marshall employee, Neil Meyers, a mechanical engineer, who had developed a lockable joint with a hinge brake.

Horton licensed the technology from Marshall and then set about applying the design concept to a new type of orthotic, a knee that automatically unlocks during the swinging phase of walking, but then is able to reengage for stability upon heel strike.

Product Outcome

Horton left Marshall with the basic design of the lockable knee joint, but still needed several years of design, development, and testing before bringing the medical device to the market.

Horton contacted Arkansas Manufacturing Solutions (AMS), operating at the time as the Arkansas Manufacturing Extension Network, a program of the Arkansas Science and Technology Authority. AMS has been instrumental in helping hundreds of Arkansas manufacturers increase sales and profits by cutting costs and improving manufacturing processes by providing technical and management assistance that improves the

Designed for patients with weak or absent quadriceps and varying degrees of knee instability, this lightweight orthosis will allow patients to regain their mobility and assist them in a more energy-efficient ambulation.

quality, productivity, and global competitiveness of state businesses. Through AMS, Horton was connected with Professor John Hebard, of the University of Arkansas at Little Rock, who helped the orthotist overcome additional design obstacles.

In total, Horton spent 7 years perfecting the design of the knee joint. The result was the Stance Control Orthotic Knee Joint (SCOKJ). Designed for patients with weak or absent quadriceps and varying degrees of knee instability,

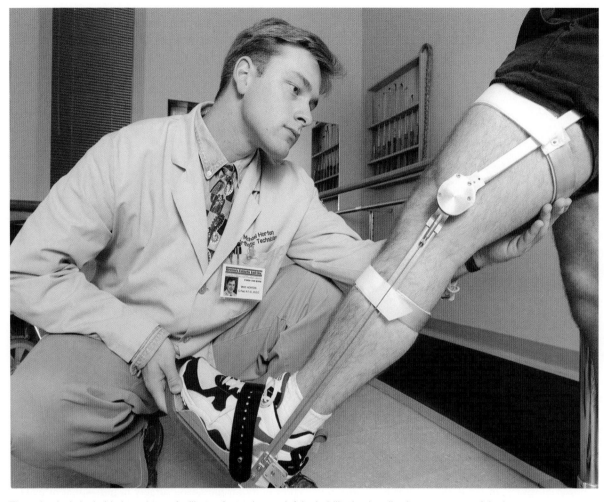

The selectively lockable knee brace facilitates faster, less painful rehabilitation by allowing movement of the knee.

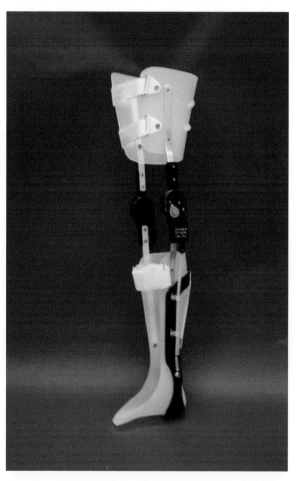

The brace may be used by a patient recovering from a knee injury when the knee cannot carry the full weight of the patient.

it is ideal for people with weak quadricep muscles due to polio, spinal cord injuries, and other conditions; such as unilateral leg paralysis. The lightweight orthosis allows patients to regain their mobility and assist them in more energy-efficient ambulation.

Much like the human knee, the selectively lockable joint operates in three distinct but complementary modes: free motion or automatic stance control, for walking, and manual lock, for standing. The device locks the knee when the heel strikes, but then releases when the heel lifts off the ground, which provides the user with a normal gait, while also providing stability while standing. The stance control feature can be triggered by weight bearing or joint motion, according to patient needs.

The SCOKJ entered the commercial realm in 2002 as a very durable option for knee orthotics and has since helped thousands of people. ❖

Stance Control Orthotic Knee™ is a trademark, and SCOKJ® is a registered trademark of Horton Technology Inc.

Robotic Joints Support Horses and Humans

Originating Technology/NASA Contribution

A rehabilitative device first featured in *Spinoff* 2003 is not only helping human patients regain the ability to walk, but is now helping our four-legged friends as well. The late James Kerley, a prominent Goddard Space Flight Center researcher, developed cable-compliant mechanisms in the 1980s to enable sounding rocket assemblies and robots to grip or join objects. In cable-compliant joints (CCJs), short segments of cable connect structural elements, allowing for six directions of movement, twisting, alignment, and energy damping.

Kerley later worked with Goddard's Wayne Eklund and Allen Crane to incorporate the cable-compliant mechanisms into a walker for human patients to support the pelvis and imitate hip joint movement.

Partnership

In June 2002, Enduro Medical Technology, of South Windsor, Connecticut, licensed NASA's cable-compliant technology and walker, and modified them into an advanced walker with a specialized, flexible harness that supports the torso. This eliminated the need for physical support from a therapist. According to Kenneth Messier, Enduro's president, the company "saw using this cable-compliant mechanism as a way to really improve and revolutionize how physical therapy is done for patients." The company designed four versions of its Secure Ambulation Module (S.A.M.), a device which provides a stable environment for patients during ambulation therapy. Enduro also introduced electronic linear actuators to give medical staff the ability to adjust and control the weight bearing of each patient, and a digital readout to record settings and track progress.

Enduro further developed the adjustable patient harness system, introducing S.A.M. in March 2003. The pelvic harness comes in various sizes and is padded with NASA-developed temper foam for comfort. The S.A.M. is

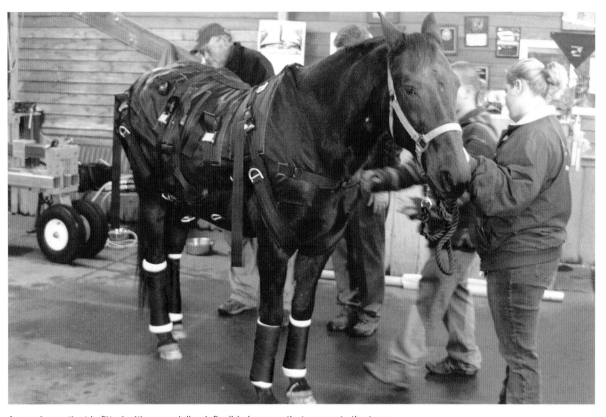

An equine patient is fitted with a specialized, flexible harness that supports the torso.

currently in use for injured veterans at Walter Reed Army Medical Center, in Washington, DC, and the Edward Hines Jr. Veterans Administration Hospital, in Chicago. Kindred Hospital, in Greensboro, North Carolina, is using an institutional S.A.M. for patients weighing up to 1,000 pounds.

Product Outcome

In response to a request from the veterinary community, Enduro engineers developed a rehabilitative device for horses using the CCJ-technology in S.A.M. Enduro secured a license from Goddard in February 2007, to develop the Enduro N.E.S.T. (NASA Equine Support Technology). Enduro believes the N.E.S.T. will revolutionize veterinary equine medicine, opening the possibility for life saving surgery in horses that otherwise may have been euthanized. Like the S.A.M., the N.E.S.T. can adjust for height and width, accommodating horses of different sizes, from smaller horses weighing 1,000 pounds to draft horses weighing 2,400 pounds.

The N.E.S.T. reduces risk both before and after surgical procedures by supporting a horse's weight. This also allows the anesthetized horse to remain upright, while traditional methods require the anesthetized horse be hoisted upside down by the legs for transportation into the surgical suite, which can cause a dangerous drop in the horse's blood pressure. After surgery, horses need to stand as soon as possible after waking, since they cannot remain lying down for extended periods. When waking from anesthesia, horses are disoriented and unstable, frequently kicking and thrashing into the padded walls of their stall while attempting to stand. This frequently leads to broken limbs, concussions, and other injuries.

The only other options for a safer surgical recovery are padded recovery stalls and recovery pools, such as the one used for the racehorse Barbaro at the University of Pennsylvania. These treatments have drawbacks and are not appropriate for all cases: for instance, recovery pools require a large number of personnel and are expensive to build and maintain, and there are only two pools currently available in the United States for equine post-surgical recovery. These specially designed pools can cost over $1 million to construct; meanwhile, the N.E.S.T. will sell for $75,000 to $90,000.

A horse recovering from surgery in the N.E.S.T. is securely positioned in a natural standing position, which reduces complications, limits additional injuries to the horse, and protects both patients and staff from injury during and after procedures. "This equipment revolutionizes how horses recover from anesthesia immediately following surgery as well as allows long-term un-weighted rehabilitation," said Messier. In the N.E.S.T., the horses remain calm, Messier explains. Conventional un-weighted therapies include underwater treadmills, rehabilitative swimming pools, and harnesses attached to permanent overhead lifts. With any water treatment, however, there is the potential for a dangerous infection if there are sutures from surgical procedures.

Traditionally, even after horses recover safely from surgery, they are still at risk of developing laminitis for months after an injury. One of most dangerous illnesses in horses, laminitis can occur when the "good" limb opposite the injured limb is forced to support too much weight. "The Enduro N.E.S.T. may help prevent the onset of this disease by selectively un-weighting different limbs," says Messier, which is possible through the use of the cable-compliant joints developed at Goddard. The unit can balance weight individually for horses' limbs instead of being weighted evenly or too heavily on one limb; Barbaro was eventually euthanized because of laminitis.

The N.E.S.T. can be used for extended periods of rehabilitation where the horse needs to stand in a controlled, secure environment. It can also be used at equine rehabilitation clinics, or brought to barns where the horses can be treated on site. Horses may even be able to reside permanently in the N.E.S.T.

Enduro continues to invest in further development and exploration of CCJ technology-based devices. A "Sit-To-Stand" version of S.A.M., also based on CCJ technology, is being used to help patients stand independently. The company has developed the S.A.M.-Y, a youth version for patients between 35 and 150 pounds and a maximum height of 5 feet 3 inches. Future Enduro plans call for the development of an adult home version of the S.A.M. ❖

The S.A.M.™ and Enduro N.E.S.T.™ are trademarks of Enduro Inc.

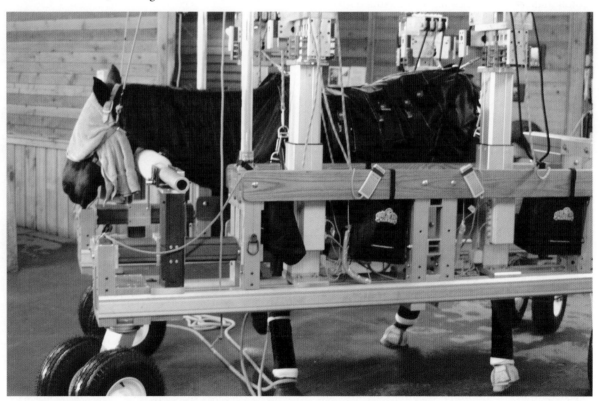

The Enduro N.E.S.T. allows horses to remain standing during anesthesia or rehabilitation.

Photorefraction Screens Millions for Vision Disorders

Originating Technology/NASA Contribution

Who would have thought that stargazing in the 1980s would lead to hundreds of thousands of schoolchildren seeing more clearly today? Collaborating with research ophthalmologists and optometrists, Marshall Space Flight Center scientists Joe Kerr and the late John Richardson adapted optics technology for eye screening methods using a process called photorefraction. Photorefraction consists of delivering a light beam into the eyes where it bends in the ocular media, hits the retina, and then reflects as an image back to a camera. A series of refinements and formal clinical studies followed their highly successful initial tests in the 1980s.

Evaluating over 5,000 subjects in field tests, Kerr and Richardson used a camera system prototype with a specifically angled telephoto lens and flash to photograph a subject's eye. They then analyzed the image, the cornea and pupil in particular, for irregular reflective patterns. Early tests of the system with 1,657 Alabama children revealed that, while only 111 failed the traditional chart test, Kerr and Richardson's screening system found 507 abnormalities.

Partnership

In 1991, NASA transferred the exclusive license for the system to the Vision Research Corporation (VRC), of Birmingham, Alabama, after Kerr sold his company to VRC. Jim Kennemer, VRC's president, says the basic technology is still the same in 2008. "What makes this work is the optics, the distance, the angles, and the flash. We retained the basic optical principles."

Also in 1991, VRC began a two-pronged marketing effort for the VisiScreen Ocular Screening System-Clinical (OSS-C): sales to pediatricians and family practitioners, and the widespread distribution of screening services to school systems and other organizations with large numbers of children. That year, the Russell Corporation

(based in Alexander City, Alabama, and now also in Atlanta) joined forces with VRC to conduct a large-scale eye-screening program in Russell's employee daycare centers. In approximately 10 percent of the children, the program identified previously unsuspected vision problems significant enough to warrant follow-up examination by an eye care professional. Because several eye conditions can worsen and even cause blindness if not caught early, there were clear benefits in continuing the screenings.

The success of this program led Russell Corporation to collaborate with Alabama Power and the Alabama State Department of Education to sponsor a program for all kindergarten students in the state. "That was the first statewide eye screening program using advanced technology in the country," Kennemer says. VRC also contributed to the growth of the nonprofit organization, Sight Savers of Alabama, to provide vision care and assistance to needy children.

VRC has since used VisiScreen to check over 3 million children in schools and daycare centers. "The NASA technology that has made our screening programs possible has truly changed the lives of hundreds of thousands of children," Kennemer says. This impact is one of the reasons the Space Foundation inducted VisiScreen into the Space Technology Hall of Fame in 2003.

A specialist from the Vision Research Corporation (right) uses the VisiScreen to photograph a patient's eyes at a specific distance and angle. Specialists at VRC later analyze the images and issue a report to the family or physician, indicating areas of possible concern.

Product Outcome

Using photorefraction, VRC's VisiScreen photographs a patient's eyes at a specific distance and angle; light enters the eye, reflects off the retina, and returns an image to the screening system. Specialists at VRC later analyze the images and issue a report to the family or physician, indicating areas of possible concern.

Although not intended to replace examination by an eye care professional, VisiScreen can highlight possible problems that a child's parents and teachers may not have noticed. The system can detect common childhood vision problems, including myopia (nearsightedness), hyperopia (farsightedness), astigmatism (corneal irregularities), strabismus (alignment errors), and cataracts, which occur in roughly 1 in every 1,000 infants.

VisiScreen tests the eye for refractive error and obstruction in the cornea or lens. The photorefractor analyzes the retinal reflexes generated by the subject's response to the flash. If the eye is properly focusing the light, as happens in a child with normal vision, a smooth, clear "red eye" image of the retina reflects evenly from the pupils. For a child who is hyperopic, a bright half-moon reflects from the top of the pupil. In the case of myopia, a crescent in the bottom half of the eye reflects more brightly than the top. Similarly, other potential problem areas reflect differently than a properly focused eye would.

The system provides several major advantages over traditional vision screening with letter or picture charts: children do not need to respond during the test, so anyone, including an infant, can be screened regardless of age or verbal ability; the process is also as quick as taking a photograph, so screeners can process large numbers of patients rapidly.

Pediatricians and family doctors in over 20 states use VisiScreen to identify possible vision problems in children, who are then referred to ophthalmologists and optometrists for diagnosis and treatment. VRC screened approximately 150,000 Alabama elementary school

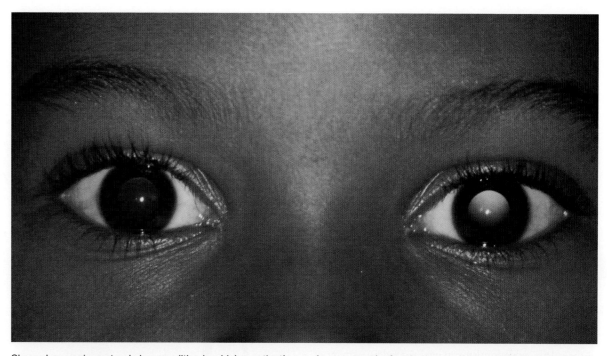

Shown here, anisometropia is a condition in which a patient's eyes have unequal refractive power. If not caught early enough, anisometropia can cause amblyopia, or lazy eye. Vision Research Corporation's screening program has found over 3,000 children with indications of anisometropia.

students during the 2007-2008 school year, and continues to offer its services across the Southeast. According to Kennemer, "Over 3,000 had indications of a difference in the power of the eyes called anisometropia, which can indicate or lead to amblyopia."

Commonly known as "lazy eye," amblyopia can cause permanent vision loss in the weaker eye if not detected and corrected early enough; "it is the leading cause of preventable blindness in children," Kennemer explains. Since its inception, VisiScreen has found amblyopic factors in over 70,000 children, or approximately 1 child in 40. "These screenings are vitally important," Kennemer says, "because if children are not treated before age 6 or 7, they may suffer permanent vision loss; in addition, amblyopia leads to 17 percent of all adult total blindness. Although

blindness is not a concern for most children in the screenings, limited vision can affect both social and educational development. A child who cannot see well is at an obvious disadvantage in the classroom, and those who fall behind early in their education are more likely to have additional problems later."

Through the efforts of VRC and VisiScreen, NASA has improved the lives of hundreds of thousands of children whose eye problems may have remained undiagnosed or otherwise untreated. VRC is planning more improvements and enhancements to VisiScreen, and soon will begin field testing a new generation of the screening system. ❖

VisiScreen® is a registered trademark of Vision Research Corporation.

Periodontal Probe Improves Exams, Alleviates Pain

Originating Technology/NASA Contribution

"Dentists," comedian Bill Cosby memorably mused, "tell you not to pick your teeth with any sharp metal object. Then you sit in their chair, and the first thing they grab is an iron hook!" Conventional periodontal probing is indeed invasive, uncomfortable for the patient, and the results can vary greatly between dentists and even for repeated measurements by the same dentist. It is a necessary procedure, though, as periodontal disease is the most common dental disease, involving the loss of teeth by the gradual destruction of ligaments that hold teeth in their sockets in the jawbone. The disease usually results from an increased concentration of bacteria in the pocket, or sulcus, between the gums and teeth. These bacteria produce acids and other byproducts, which enlarge the sulcus by eroding the gums and the periodontal ligaments.

The sulcus normally has a depth of 1 to 2 millimeters, but in patients with early stages of periodontal disease, it has a depth of 3 to 5 millimeters. By measuring the depth of the sulcus, periodontists can have a good assessment of the disease's progress. Presently, there are no reliable clinical indicators of periodontal disease activity, and the best available diagnostic aid, periodontal probing, can only measure what has already been lost. A method for detecting small increments of periodontal ligament breakdown would permit earlier diagnosis and intervention with less costly and time-consuming therapy, while overcoming the problems associated with conventional probing.

The painful, conventional method for probing may be destined for the archives of dental history, thanks to the development of ultrasound probing technologies. The roots of ultrasound probes are in an ultrasound-based time-of-flight technique routinely used to measure material thickness and length in the Nondestructive Evaluation Sciences Laboratory at Langley Research Center. The primary applications of that technology have been for corrosion detection and bolt tension measurements (*Spinoff*

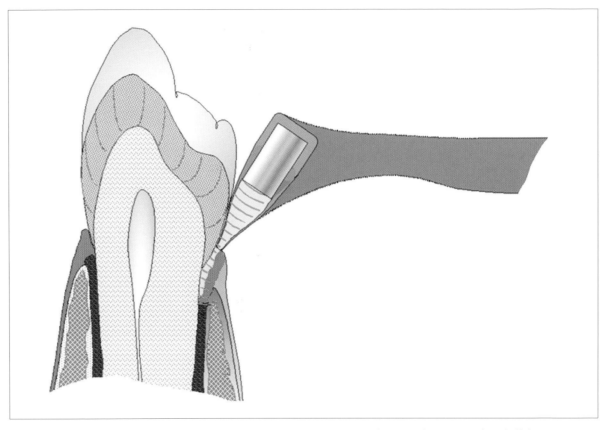

The USProbe device rides along the gum line, sending a signal down the pocket between the gums and teeth. Echoes are recorded by an ultrasound transducer and then analyzed by a computer system. A foot pedal is used to start the flow of water, read data, and then send the information to a computer where it is fed into a charting software program.

2005). This ultrasound measurement system was adapted to the Periodontal Structures Mapping System, invented at Langley by John A. Companion, under the supervision of Dr. Joseph S. Heyman. Support of the research and development that led to this invention was provided by NASA's Technology Applications Engineering Program and by the Naval Institute for Dental and Biomedical Research, in Great Lakes, Illinois. In fact, a request from the U.S. Navy spurred the development of the tool: A sailor on a submarine had to be airlifted 1 month into a 6-month tour due to a life threatening case of periodontal disease, costing the Navy millions of dollars as the mission had to be abandoned.

Partnership

Patented as the Ultrasonagraphic Probe (USProbe) in May 1998, Visual Programs Inc., of Richmond, Virginia, obtained an exclusive license for the system in January 2000. According to John Senn, of Visual Programs, the new device may be one of the major steps forward in the

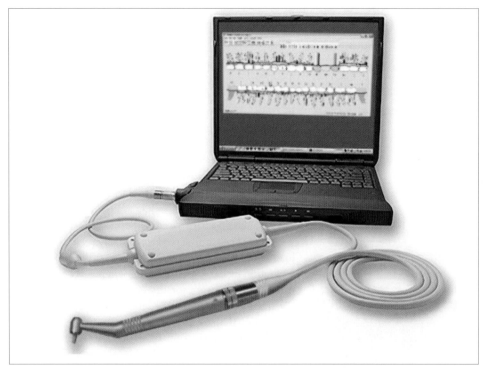

The first PC-based automated probe on the market, USProbe integrates with most charting systems and allows one person to conduct a periodontal probing diagnostic exam while recording the results.

Visual Programs developed the USProbe as the next-generation, state-of-the-art diagnostic tool for detecting and characterizing periodontal disease. The USProbe automatically detects, maps, and diagnoses problem areas by integrating diagnostic medical ultrasound techniques with advanced artificial intelligence. Visual Programs expects it will quickly become the industry standard technique, replacing the current uncomfortable and invasive techniques. NASA and Visual Programs are proud to contribute technology that will increase the number of healthy smiles and decrease the number of grimaces produced by their maintenance. ❖

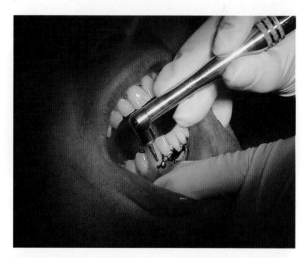

Much more comfortable than conventional probing, USProbe uses a slight flow of water to ensure coupling of ultrasound in and out of tissues.

battle against periodontal disease. "The probe should be the next major piece of dental equipment. By using the new technology, dentists and hygienists will be able to perform exams earlier and may detect periodontal disease while the teeth can still be saved." According to Jack Singer, the president of Visual Programs, "The name NASA has opened many doors for us that may not have been opened otherwise. It gives credibility to a new concept that otherwise might not have been accepted."

Product Outcome

The USProbe mapping system is a noninvasive tool to make and record differential measurements of a patient's periodontal ligaments relative to a fixed point, the boundary between the crown and root of a tooth, called the cemento-enamel junction (CEJ). The mapping system

uses ultrasound to detect the top of the ligaments at various points around each tooth, and uses either ultrasound or an optical method to find the CEJ at the same points. The depth of the sulcus is calculated as the difference between these two points.

The probe used in the mouth to send and receive ultrasound signals is very small, and additional instrumentation is contained within a standard personal computer, allowing the entire measurement to be computerized. In addition, manual charting of pocket depth will be eliminated, since the data will be automatically transmitted to the computer. In addition to solving the problems associated with conventional probing, the USProbe may also provide information on the condition of the gum tissue and the quality and extent of the bond to the tooth surface.

Magnetic Separator Enhances Treatment Possibilities

Payload Specialist John Glenn worked with Techshot's ADSEP hardware aboard Space Shuttle Discovery on STS-95.

Originating Technology/NASA Contribution

Since the earliest missions in space, NASA specialists have performed experiments in low gravity. Protein crystal growth, cell and tissue cultures, and separation technologies such as electrophoresis and magnetophoresis have been studied on Apollo 14, Apollo 16, STS-107, and many other missions.

Electrophoresis and magnetophoresis, respectively, are processes that separate substances based on the electrical charge and magnetic field of a molecule or particle. Electrophoresis has been studied on over a dozen space shuttle flights, leading to developments in electrokinetics, which analyzes the effects of electric fields on mass transport (atoms, molecules, and particles) in fluids. Further studies in microgravity will continue to improve these techniques, which researchers use to extract cells for various medical treatments and research.

Partnership

As part of the NASA Shuttle Student Involvement Program, John Vellinger designed an experiment to study the effect of low gravity on chicken embryos. After his freshman year at Purdue University in 1985, Vellinger partnered with KFC Corporation staff engineer, Mark Deuser, who helped Vellinger design the "Chix in Space" experiment for two space shuttle missions. Their work together laid the foundation for creating Space Hardware Optimization Technology Inc. (SHOT) in 1988, after acquiring a NASA contract for space flight hardware development.

Since 1988, four NASA centers—Marshall Space Flight Center, Glenn Research Center, Johnson Space Center, and Ames Research Center—have issued over 25 **Small Business Innovation Research (SBIR)** contracts to the Greenville, Indiana-based company, which has since changed its name to Techshot Inc. For its first 13 years, Techshot served exclusively as a NASA payload company, specializing in space hardware and later, separation technologies. Techshot engineers designed and integrated hardware for three suborbital rocket flights, seven space shuttle missions, and several payloads for the International Space Station.

Working with the Consortium for Materials Development in Space at the University of Alabama in Huntsville, Techshot won a contract to develop separation technologies for space application. Separation technologies involve the analysis and purification of proteins or "markers" which reveal information about diseases and the body. With the addition of Paul Todd as chief scientist in 2000, Techshot realized the value of terrestrial applications of separation technology, and explored experimental possibilities with Marshall and Johnson.

Product Outcome

Accustomed to working with the exacting requirements for space payloads, Techshot took a logical step in commercializing separation technology for the equally strict medical field, beginning with an organic separator developed for the Organic Separation Experiment that first flew on STS-57. Next, the company developed advanced separators based on the Advanced

Space Experiment Processor (ADSEP), operated in low gravity without using a centrifugal field. These technologies provide biphasic, electrophoretic, and magnetic separation capabilities.

During early development of these separators, Techshot found the number of cells collected was insufficient for the needs of the experiment, for example, in stem cell transplants. To remedy this, Techshot purchased the license for a magnetic separation technology from the Cleveland Clinic Foundation, already shown to produce enough cells for a stem cell transplant. The resulting system—Magsort—is a multistage electromagnetic separator for purifying cells and magnetic particles for a variety of research and medical uses, including stem cell research and cancer treatment, in a much more refined manner than previously possible.

A laboratory apparatus for separating particles (especially biological cells), Magsort separates cells based on their magnetic susceptibility and magnetophoretic mobility. Magsort refines a sample population of particles by categorizing multiple portions of the sample and separating substances into individual parts, instead of as entire sets. As Todd explains, "If you have a mixture of cells—A, B, C, and D—Magsort will separate and collect A and B and C and D and put them separately in the hands of a research investigator . . . . Other existing methods would sort out only B, for example."

The cells of interest might only be present in low numbers, so refined separation techniques are necessary for effective research. In its magnetophoretic process, the Magsort creates magnetic particles that are smaller than the cell but react only to one certain cell type. The hematological stem cells, for example, have molecules on their surface (clusters of differentiation 34) that can react with an antibody and then attach to a magnetic particle. If the mixture is then placed in a magnetic field, only the stem cells will be drawn to the magnet.

Magsort includes an electronic unit coupled to a built-in computer and a processing unit that includes horizontal

Techshot and IKOTech licensed and commercialized this Quadruple Magnetic Sorter, which uses magnetophoresis for cancer research and stem cell processes.

upper and lower plates, a plate rotation system, a graded series of capture magnets above the upper plate, and a wheel for sequencing the small permanent capture magnets. The bolted plates rotate, and an interface between them acts as a seal for separating fluids. Up to 15 upper cuvette stations allow for fraction collection; a stepping motor drives the rotation system, causing the upper plate to rotate for fractional sample collection. User interface software displays the status of the various components on a small organic light-emitting diode (LED) monitor: the translating electromagnet, capture magnet, and the rotation of the upper plate.

Techshot is currently developing other separation products for the medical field. A Techshot spinoff company, IKOTech LLC, is commercializing a version of the Magsort called the Quadruple Magnetic Sorter; it is used primarily for stem cell processes and cancer research and treatment, including the detection of rare cancer cells and the removal of undesired cells from bone marrow transplants. Another model is being developed for type 1 diabetes pancreas islet transplants. Todd explains, "Right now, it takes two or three donor organs to treat one person. Our process greatly increases the yield." With enough healthy (transplanted) beta cells of their own, patients' symptoms could vanish.

Techshot credits the success of their separation technologies with their years of experience with NASA, and consider the company itself a spinoff. Rich Boling, vice president of corporate advancement, claims that Techshot's years of experience in payloads made their commercial separators possible, as their capabilities as a company were "refined in the crucible of human space flight." ❖

NASA-derived technologies enhance ways that we travel. The benefits featured in this section:

- Deliver electric vehicles to market
- Increase helicopter stability
- Revolutionize truck design
- Ease and speed prototyping
- Perform complex design analysis

Lithium Battery Power Delivers Electric Vehicles to Market

Originating Technology/NASA Contribution

As increased energy efficiency, and particularly fuel efficiency, becomes a greater concern, hybrid and electric vehicles gain greater prominence in the market. Electric vehicles (EVs), in particular, provide an attractive option as they produce no emissions during operation, isolating any potential emissions and effluents in the manufacturing and energy-generation streams.

The necessary energy stores to support a shift to EVs already exist, as utilities constructed to address peak demands have off-peak surpluses sufficient to charge about 180 million plug-in hybrid or all-electric cars. According to a report from the U.S. Department of Energy's Pacific Northwest National Laboratory, there is enough excess generating capacity during the night and morning to allow more than 80 percent of today's vehicles to make the average daily commute solely using this electricity. Effective energy management sees its ultimate realization in the vehicle-to-grid (V2G) concept, in which plug-in hybrid and electric vehicles can be used to balance energy demand and consumption. In a V2G system, millions of automotive batteries could absorb excess power generated, and release it back into the grid at times of insufficient supply. With a several kilowatt-hour storage capacity per vehicle, millions of operational plug-ins could act as a safety net for the power grid, supplying backup power in an outage, with the vehicle owners credited for power returned to the grid. This smoothing of excess and deficiency in the power grid would also help stabilize intermittent sources of energy such as wind power and make them more viable alternatives.

Historically, the primary obstacles to the widespread application of EVs were lack of infrastructure development and a lack of sufficiently robust battery technologies to consistently power vehicles for an extended duration and at performance levels suitable to a modern urban environment. Technology may at last have caught up with the need, and rising petroleum prices

Hybrid Technologies worked with Kennedy Space Center in the testing and development of its line of electric vehicles.

are encouraging more and more consumers to consider electric and hybrid vehicles. In addition, a study by the U.S. Department of Transportation has indicated that plug-in cars capable of 50 miles per day would meet the needs of 80 percent of the American driving public, the average daily commuters.

NASA has taken a keen interest in battery-powered vehicles, and is encouraging their continued development. The "NASA Official Fleet Management Handbook," regarding the use of alternative fueled vehicles, states: "Ideally, all Centers should have on-site alternative fuel facilities . . . . Centers are encouraged to use NEVs [Neighborhood Electric Vehicles] to fill inventory requirements where feasible."

Partnership

Hybrid Technologies Inc., a manufacturer and marketer of lithium-ion battery-EVs, based in Las Vegas, Nevada, and with research and manufacturing facilities in Mooresville, North Carolina, entered into a Space Act Agreement with Kennedy Space Center to determine the utility of lithium-powered fleet vehicles. Under this agreement, the company supplied a fleet of cars for the engineers at Kennedy to test. In return for the engineering expertise supplied by the NASA employees, the Center was given the opportunity to use the zero-emission vehicles for transportation around the Kennedy campus. NASA contributed engineering expertise for the cars' advanced battery management system, and vehicles selected for use in the Kennedy fleet included the Hybrid PT Cruiser, lithium smart fortwo, and a high-performance all-terrain vehicle.

The vehicles were powered by Ballard Power Systems' 312V 67 MS electric drive system, which has a 32kW continuous rating and delivers a peak power of 67kW, with torque of 190 Nm (140 lb-ft). Hybrid Technologies selected this motor based on its proven track record and excellent power-to-weight ratio. The electric PT Cruisers have a top speed in excess of 80 miles per hour and a range of 120 miles. Charge time is 6-8 hours with either 110-120 V or 220-240V, and the lithium-ion battery pack has a cycle life of more than 1,500 charges.

In addition to the vehicles supplied to NASA, the company provided a fleet of lithium-ion battery-powered vehicles for use by the U.S. Environmental Protection Agency and the U.S. Navy.

Product Outcome

Hybrid Technologies deployed the first all-electric taxi in New York City and has begun demonstrating smart fortwo conversions like the ones used at Kennedy. The company also delivered an additional two PT Cruiser-based electric taxis and an electric Chrysler Town &

Country minivan to the city of Sacramento for use by a private para-transit nonprofit organization. Most recently, Hybrid Technologies has produced an EV version of the popular MINI Cooper, which debuted in the December 2007 Sam's Club catalog. The EV MINI Cooper proudly displays its NASA heritage, sharing the STS-128 designation with an upcoming Space Shuttle Endeavor mission. It boasts a range of 120 miles at 75 miles per hour, and is driven by a 40kW electric motor and powered by a 30kWh battery pack. The appeal of the electric MINI is strong and widespread, and Hybrid Technologies conversions have already attracted celebrity fans.

Also available from Sam's Club, the 2007 Hybrid Technologies lithium-powered smart fortwo EV (also available as a limited edition STS-118 smart fortwo) has an estimated range of 150 miles, a top speed over 70 mph, and takes only 4 hours to charge at 220 volts. There are two electric motors that can be used in the vehicle, one from Ballard and one from Siemens VDO. The lithium polymer battery pack comes from Kokam America Inc., and the battery management system is Hybrid Technologies' own. As an introductory offer, Sam's Club included a behind-the-scenes trip to Kennedy and attendance at a space shuttle launch, with purchase of one of the EVs. When asked about the availability of amenities such as air conditioning and heating, comforts not always incorporated into EV conversions, Richard Griffiths, Strategic Relations for Hybrid Technologies, stated "The [smart fortwo EV] has absolutely every option, every feature that a regular, production smart car has." Griffiths estimated the extra amenities consume about 5 percent of the vehicle's battery capacity. "We're offering the fully electric smart car to Sam's Club members as it represents the latest in advanced lithium technology . . . . This limited edition STS-118 smart car will be the perfect addition for car collectors or the environmentalist wanting to make a difference by driving a zero emissions vehicle." In addition to the MINI Cooper and smart fortwo

The EV MINI Cooper proudly displays its NASA heritage and boasts a range of 120 miles at 75 miles per hour.

conversions, Hybrid Technologies offers PT Cruiser and Chrysler Crossfire EV conversions.

Even more impressive than its line of conversions, Hybrid Technologies now also offers a series of purpose-built lithium electric vehicles dubbed the LiV series. The LiV series is designed from the ground up at Hybrid Technologies' Mooresville plant. The LiV Wise is aimed at the urban and commuter environments, and is larger and offers more interior space than the smart car, the conversion of which is called the LiV Dash. Hybrid Technologies has rounded out the LiV line with custom motorcycles, utility vehicles, mobility scooters, bicycles, and even a military vehicle. Hybrid Technologies plans to offer these vehicles to the U.S. market on a wider scale by 2009, and is especially focused on developing a system that will seamlessly integrate LiV Wise cars in small markets by 2009 and mass markets by 2010. ❖

Hybrid Technologies offers the LiV series of purpose-built lithium electric vehicles, which also includes custom motorcycles, utility vehicles, and mobility scooters.

LiV™, Wise™, and Dash™ are trademarks of Hybrid Technologies Inc.

MINI Cooper® is a registered trademark of Bayerische Motoren Werke AG.

PT Cruiser®, Town & Country®, and Crossfire® are registered trademarks of Chrysler Corporation.

smart® and fortwo® are registered trademarks of Daimler AG.

Advanced Control System Increases Helicopter Safety

Originating Technology/NASA Contribution

For over 30 years, NASA and U.S. Army engineers have worked together at Ames Research Center to make rotorcraft fly more quickly, quietly, and safely in all kinds of weather. Development of new technologies for both military and civil helicopters, tiltrotor aircraft, and other advanced rotary-wing aircraft has engaged disparate parties from all corners of the rotorcraft industry, the U.S. Department of Defense, and other government agencies. These programs have focused on all manner of helicopter components:

- Cockpit controls: Cockpit layout and design can profoundly affect the ease or difficulty of piloting a rotorcraft.

- Handling and performance: NASA and Army experts design flight control systems which make helicopters and other rotorcraft easier to fly using a full-motion simulator and actual aircraft.

- Noise: Most rotorcraft noise results from vibrating parts and the interaction of air vortices shed from the tips of the rotors. Researchers use wind tunnels to investigate ways to reduce noise.

- Speed and performance: Airflow around the fuselage and moving rotor blades is very complex. These complexities limit the helicopter's speed in moving in different directions. Ames researchers use wind tunnels and computers to investigate ways to improve the airflow.

Particularly focused on safe rotorcraft operation, NASA's Safe All-Weather Flight Operations for Rotorcraft (SAFOR) element of the Rotorcraft Research and Technology Base Program was specifically tasked with improving the safety of civil helicopter operations. SAFOR ran from 1999 through 2002 and focused on improving drive systems technology, flight control and guidance technology, and situational awareness and information display technologies.

Preset attitude retention limits in the HeliSAS prevent inadvertent cyclic inputs that could result in an unusual attitude.

The drive systems element sought to reduce the frequency and consequences of main and tail rotor, transmission, drive, clutch, gearbox, and drive system failures by improving reliability of drive systems. The flight control segment included work to reduce the frequency and severity of accidents due to loss of control, high workload, and exceeding vehicle limits. The situational awareness and information displays unit pursued reduced frequency and severity of accidents due to pilot error, inexperience, poor judgment, lack of situational awareness, and inadequate preparation. SAFOR led to many improvements in helicopter design and operation, some of which have already reached the commercial market.

Partnership

Hoh Aeronautics Inc. (HAI), of Lomita, California, was founded in 1988 and is dedicated to the analysis and development of conventional and advanced flight control systems and displays for fixed and rotary wing aircraft. HAI engineers also evaluate and develop handling qualities criteria, piloted simulations and flight-test programs, and computer-based training programs.

With support and funding from a Phase II NASA **Small Business Innovation Research (SBIR)** project from Ames, HAI produced a low-cost, lightweight, attitude-command-attitude-hold stability augmentation system (SAS) for use in civil helicopters and unmanned aerial vehicles (UAVs). The primary advantage of the SAS is that, by increasing helicopter stability and allowing hands-free operation of the aircraft, the system helps the pilot to accomplish divided attention tasks. SAS improves helicopter dynamics and enhances safety in low-speed and hovering maneuvers in degraded visual environments, and for Instrument Flight Rules (IFR) operations in forward flight. As opposed to Visual Flight Rules (VFR), IFR operation of the vehicle references only the instruments and Air Traffic Control, allowing operation in conditions that obscure the pilot's view; most commercial air traffic operates exclusively under IFR.

The prototype helicopter autopilot/stability augmentation system, dubbed HeliSAS, weighed 12 pounds, significantly less than comparable systems, which can weigh over 50 pounds. HeliSAS proved its superior performance in over 160 hours of flight testing and demonstrations in a Robinson R44 Raven helicopter, one of the most popular commercial helicopters and a particular favorite of news broadcasting and police operations. The HeliSAS reduced pilot workload and increased safety by allowing hands-off flight, and as an added bonus, the system cost significantly less than current systems that perform the same functions.

By offering significant stability and control improvements in a low-cost/lightweight system, HeliSAS promises many benefits in space, military, and civilian aviation applications, including:

- Improving the stability of light helicopters at an affordable cost without excessive weight penalty

- Increasing feasibility of low-cost UAVs

- Potentially developing dual-role, low-cost utility helicopter/UAVs, which can be flown with or without a pilot

Product Outcome

HAI developed HeliSAS into a superior stability augmentation system for light helicopters. With the push of a button, the HeliSAS converts the R44 Raven from an unstable aircraft with very light stick forces to a highly stable platform with enhanced control feel that provides force feedback to the pilot—in effect, HeliSAS makes the R44 feel like a much larger, more stable helicopter. With the system engaged, it is possible for the pilot to remove his or her hand from the cyclic to fold charts or perform other cockpit duties, and the R44 has been demonstrated to hold attitude indefinitely with the HeliSAS engaged. A full autopilot option has been added, including altitude hold, heading select/hold, VHF Omni-directional Radio Range Localizer (VOR/LOC) track, Instrument Landing System (ILS) track, and Global Positioning System (GPS) steering.

Shawn Coyle, an instructor at the National Test Pilot School, a not-for-profit educational institute incorporated in California, and former Civil Aviation Authority certification pilot in Canada, conducted a flight evaluation of the HeliSAS, and the system has been featured in Helicopter World magazine's "North American Special Report 2004" (published in the United Kingdom).

Chelton Flight Systems, of Boise, Idaho, negotiated with HAI to develop, market, and manufacture HAI's HeliSAS autopilot system, and the product is now available as the Chelton HeliSAS Digital Helicopter Autopilot. ❖

HeliSAS™ is a trademark of Hoh Aeronautics Inc.

HeliSAS provides unmatched stability and ease of handling, including "hands near" operation while maintaining heading or navigation course and attitude—even during turbulent conditions.

Aerodynamics Research Revolutionizes Truck Design

Originating Technology/NASA Contribution

The last 35 years have seen a sea change in the design of trucks on America's highways, reflecting extensive research into vehicle aerodynamics and fluid dynamics conducted by NASA engineers. Thanks to the ingenuity of a Dryden Flight Research Center researcher bicycling through the California desert and a team of engineers in Virginia, the shape of rigs and recreational vehicles (RVs) today owes as much to the skies as it does the open road.

Bicyclists, motorcyclists, and even pedestrians feel a push and pull of air as large trucks pass. The larger a vehicle is and the faster it moves, the more air it pushes ahead. For a large truck, this can mean a particularly large surface moving a large quantity of air at a high velocity—its blunt face acting like a fast-moving bulldozer, creating a zone of high pressure. The displaced air must go somewhere, spilling around the cab into swirling vortices. The air traveling along the side moves unevenly, adhering and breaking away, and sometimes dissipating into the surrounding air. At the end of the cab or trailer, the opposite effect of the high-pressure zone at the front develops; the airflow is confronted with an abrupt turn that it cannot negotiate, and a low-pressure zone develops.

The high pressure up front, the turbid air alongside and under the vehicle, and the low pressure at the back all combine to generate considerable aerodynamic drag. A study published in Automotive Engineering in August 1975 found that a tractor trailer unit moving at 55 miles per hour displaced as much as 18 tons of air for every mile traveled. In such cases, roughly half of the truck's horsepower is needed just to overcome aerodynamic drag.

In 1973, Edwin J. Saltzman, Dryden aerospace engineer and bicyclist, noticed the push and pull of large trucks at highway speeds while riding to work. As a tractor trailer overtook him, he first felt the bow wave of air pushing him slightly away from the road and toward the sagebrush; as the truck swept past, its wake had the opposite effect, drawing him toward the road and even causing both rider and bicycle to lean toward the lane. Saltzman mused about ways to mitigate the bow wave and trailing partial vacuum, and resolved to help trucks glide through air instead of push through it, and, in the process, decrease drag and increase fuel efficiency. NASA colleagues at Dryden were working on the effects of drag and wind resistance on different kinds of aircraft and the early space shuttle designs, so they transferred their considerable knowledge to the design of large trucks.

The first formal experiment involved a Ford van retired from delivery duties at Dryden. Mechanics attached an external frame which was then covered with sheet aluminum to give the van flat sides all around and 90-degree angles at all corners. The vehicle looked like an aluminum shoebox on wheels, simulating the cruder motor homes of

Dryden engineers modified a retired delivery van to test aerodynamic drag, first boxing the van with aluminum sheets at 90-degree angles, and then rounding the sides and fashioning a boat-tail rear.

A cab over engine tractor trailer was leased by Dryden, tested, and modified to reduce aerodynamic drag.

the period. The Dryden engineers measured the vehicle's baseline drag and then set about modifying the shape of the van: First rounding the front vertical corners, then the bottom and top edges of the front, then the edges of the aft end, and finally sealing the entire underbody of the van including the wheel wells, with tests run after each modification. Rounding all four front edges yielded a 52-percent drag reduction, while sealing the bottom of the vehicle gained another 7 percent. The engineers esti-mated the potential gain in fuel economy to be between 15 and 25 percent at highway speeds.

During the following decade, Dryden researchers conducted numerous tests to determine which adjust-ments in the shape of trucks reduced aerodynamic drag and improved efficiency. The team leased and modified a cab over engine (COE) tractor trailer, the dominant cab design of the time, from a Southern California firm. Modifications included rounding the corners and edges of the box-shaped cab with sheet metal, placing a smooth fairing on the cab's roof, and extending the sides back to the trailer.

Rounding the vertical corners on the front and rear of the cab reduced drag by 40 percent while decreasing internal volume by only 1.3 percent. Likewise, rounding the vertical and horizontal corners cut drag by 54 percent, with a 3-percent loss of internal volume. Closing the gap between the cab and the trailer realized a significant reduc-tion in drag and 20 to 25 percent less fuel consumption. A second group of tests added a faired underbody and a boat tail, the latter feature resulting in drag reduction of about 15 percent. Assuming annual mileage of 100,000 driven by an independent trucker, these drag reductions would translate to fuel savings of as much as 6,829 gallons per year.

On the other coast from Saltzman and his Dryden team, Dr. John C. Lin and Floyd G. Howard of Langley Research Center with Dr. Gregory V. Selby of Old Dominion University, Norfolk, Virginia, conducted a series of research projects in the late 1980s and early 1990s focusing on controlling drag and the flow of air around a body. One study conducted in 1989, "Turbulent Flow Separation Control," explored controlling airflow—flow separation—to decrease energy expenditure and weight in airfoils, inlets, and diffusers and improve aircraft control and decrease drag. The study employed vortex generators, aerodynamic surfaces protruding from a body that draw faster moving air to the surface of the vehicle and disrupt the slower moving boundary layer air around a vehicle, the use of which can be traced back to research conducted by the National Advisory Committee for Aeronautics (NASA's forebear) in the 1950s. The generated vortices "energize" the slower-moving boundary layer and thereby reduce drag and, in aircraft applications, increase lift.

Subsequent studies in 1990 and 1991 continued vortex-generator research with an exploration of various active and passive methods for controlling two-dimensional separated flow. These studies quantified and characterized the behavior and performance of a variety of large-eddy breakup devices for turbulent flow separation control.

Partnership

Answering the charge given by the U.S. Congress in the National Aeronautics and Space Act of 1958 to disseminate newfound technologies and discoveries to the public, NASA makes the results of its research and expertise of its scientists and engineers available through a variety of means. Sponsored by the Innovative Partnerships Program, these include published studies, NASA outreach, the Small Business Innovation Research and Small Business Technology Transfer programs, technology transfer offices at each NASA field center, and the Space Alliance Technology Outreach Program (SATOP).

The aerodynamics studies at Dryden have been made publicly available, and Aeroserve Technologies Ltd., of Ottawa, Canada, with its marketing arm, Airtab LLC, in Loveland, Colorado, applied these studies, the aerodynamic work at Langley, and the patented Wheeler vortex generator to the development of the Airtab vortex generator; designed to reduce drag and improve vehicle stability and fuel economy. Of the devices tested, the Wheeler showed the least parasitic drag, and Aeroserve optimized the Wheeler design for ease of installation and application to any vehicle.

Product Outcome

The Surface Transportation Assistance Act of 1982 required states to permit trucks with trailers as long as 48 feet on both interstate and intrastate highways; the previous length limit of 55 feet had applied to the tractor and trailer together. As the previous regulation made the COE tractor a dominant choice, owing to its decreased length regardless of aerodynamic or fuel efficiency shortcomings, the new regulations opened the door for a renaissance of the "conventional" cab. While COE designs place the cab directly above the engine, minimizing length and producing a cube-like tractor, conventional truck designs place the engine ahead of the cab. Though longer as a result, a protruding nose offers truck designers an inherently more aerodynamic shape from which to work. In 1982, COE trucks constituted over 65 percent of the market for the Peterbilt Motors Company, with similar numbers for other manufacturers; the cab-over design represented only 1 percent of sales for Peterbilt by 2004.

Streamlined cabs and fairings are now a common sight on our highways, and the once-prominent cab-over design has been abandoned in virtually all applications except small-capacity urban-oriented trucks where length remains a premium. The modifications tried by the engineers at Dryden were adopted by the truck manufacturers, as the same principles the NASA engineers demonstrated with COE trucks applied to conventionals. In addition, the cargo boxes of most delivery trucks today have rounded corners and edges, a direct application of the research conducted at Dryden on the "shoebox."

Today's trailers, on the other hand, are little changed from the last few decades. For livestock haulers, a key factor is that individual farmers have been the predominant owners of trailers, and these owners are difficult to convince about the costs of redesign versus the savings of superior aerodynamics. However, more and more livestock trailers are sporting boat-tail designs that ease the flow of air past the end of the trailer and minimize the low-pressure wake. Conventional trailer manufacturers have resisted change more so than others, in part because the aft end of such a trailer needs to be easy to manipulate at loading docks, where the optimal shape for superior aerodynamics—the boat tail—is impractical.

Airtabs create controlled, counter-rotating vortices to bridge the gap between tractor and trailer or control airflow past the rear of the vehicle.

Likewise, the gap between the cab and the trailer can create a significant amount of drag as air swirls in the space between. Two conventional means to address this issue are problematic: Adding side extenders (to decrease the exposed gap) is expensive and might impede maneuverability; moving the fifth wheel forward (to shorten the gap) places more weight on the steering axle—which is legally regulated and limited—and reduces maneuverability while increasing driver effort and wear on steering tires and steering gear.

Addressing both of these dilemmas, Aeroserve's Airtabs garner the benefits of the airflow found in a boat-tail design with the practicality of a squared-off end for loading and unloading, and see additional applicability smoothing the airflow between cab and trailer. Airtab vortex generators create a controlled vortex to reduce truck and trailer wind resistance and aerodynamic drag. Each Airtab produces two counter-rotating vortices of air, each approximately four to five times the height of the Airtab and several feet in length, that smoothly bridge the gap between tractor and trailer or control airflow past the rear of the vehicle. Airtabs thus allow an operator to set the fifth wheel to the optimum position without incurring extra drag or steering gear wear penalties and gain some of the aerodynamic benefit of side extenders.

At the back of a trailer, box van, or RV, Airtabs radically alter the airflow to reduce drag in two ways: Shifting the airflow pattern from vertical to horizontal to eliminate large eddies, and smoothing the airflow to artificially simulate a tapered rear of the vehicle. In fact, Airtabs have been shown effective on any vehicle with more than a 30-degree slope to the rear; the potential benefits stretch across vehicular applications and could thus benefit a considerable number of vehicles.

Smoothing the airflow results in markedly improved fuel economy without compromise to design utility, and additional benefits have been realized as well. The vortex generation reduces spray; users have reported improved rear and side view in wet or snowy weather, increasing safety and offering a clearer view of surrounding vehicles. Also, because Airtabs alter the airflow around the rear of a vehicle, the accumulation of road grime is reduced, keeping tail lights and reflectors clean and allowing less snow to build up, a significant safety benefit in foul weather. Less accumulation of road grime also means advertising and safety information on the back of a vehicle remains visible.

Perhaps most importantly, drivers of vehicles fitted with Airtabs have reported improved stability and handling and dramatically reduced fishtailing of trailers—an effect where the trailer sways or slides from side to side independent of the tractor, potentially causing catastrophic loss of control—effects that are especially important with the double trailers found in North America and the famous quad-trailer "road trains" in Australia. Increased stability also means that the trailer does not scrub on the sides of the road as much, increasing the life of tires. Drivers also report better handling when being passed in the same direction by other large vehicles.

Cummins Rocky Mountain LLC, a diesel engine and generator wholesale and distribution company in Broomfield, Colorado, recognized these benefits and agreed to promote and sell Airtabs after internal testing and customer feedback indicated that Airtabs brought immediate safety and fuel economy benefits when running equipment at highway speeds. The company noted additional benefits included ease of installation, minimal maintenance, and low price.

As more NASA research and development is adapted and introduced to the market by companies like Aeroserve, the vehicles populating our highways and interstates will likewise continue to evolve. Practical solutions to aerodynamic challenges, exemplified by the Airtab, offer increased stability, safety, and economy to airborne and surface vehicles alike, and NASA is proud to contribute tangible and current benefits to both fields of transport and travel. ❖

Effective on any vehicle with more than a 30-degree slope to the rear, Airtabs see many applications in addition to trucks.

Engineering Models Ease and Speed Prototyping

Originating Technology/NASA Contribution

NASA astronauts plan to return to the Moon as early as 2015 and establish a lunar base, from which 6-month flights to Mars would be launched by 2030. Essential to this plan is the Ares launch vehicle, NASA's next-generation spacecraft that will, in various iterations, be responsible for transporting all equipment and personnel to the Moon, Mars, and beyond for the foreseeable future.

The Ares launch vehicle is powered by the J-2X propulsion system, with what will be the world's largest rocket nozzles. One of the conditions that engineers carefully consider in designing rocket nozzles—particularly large ones—is called separation phenomenon, which occurs when outside ambient air is sucked into the nozzle rim by the relatively low pressures of rapidly expanding exhaust gasses. This separation of exhaust gasses from the side-wall imparts large asymmetric transverse loads on the nozzle, deforming the shape and thus perturbing exhaust flow to cause even greater separation. The resulting interaction can potentially crack the nozzle or break actuator arms that control thrust direction.

Side-wall loads are extremely difficult to measure directly, and, until now, techniques were not available for accurately predicting the magnitude and frequency of the loads. NASA researchers studied separation phenomenon in scale-model rocket nozzles, seeking to use measured vibration on these nozzle replicas to calculate the unknown force causing the vibrations. Key to this approach was the creation of a computer model accurately representing the nozzle as well as the test cell.

Partnership

System-response models developed by LMS International NV were used to calculate side-wall loads on the J-2X nozzles. LMS is a Belgium-based company founded in 1980 with over 30 offices around the world, which acts as an engineering innovation partner for com-

LMS system-response models were used to calculate side-wall loads for the J-2X rocket nozzles, easing NASA's development and offering LMS engineers a unique opportunity to gather data.

panies in automotive, aerospace, and other advanced manufacturing industries. LMS works with customers to improve process efficiency and product quality by offering a unique combination of virtual simulation software, testing systems, and engineering services.

LMS Virtual.Lab, an integrated suite of simulation software, developed the system-response models based on modal data on nozzle replicas from LMS Test.Lab, a software solution for test-based engineering combining high-speed multichannel data acquisition with a suite of integrated testing, analysis, and report-generation tools. Tests were conducted by the Marshall Space Flight Center Structural Dynamics Test Branch, which uses LMS Test.Lab in modal testing for a wide range of projects.

The close integration between LMS Test.Lab and LMS Virtual.Lab means data is readily available without file conversions, which often fail to fully represent critical data, such as frequency response functions (FRFs). Utilizing test data in combination with modeling and predictive tools in this type of hybrid approach will enable engineers to more accurately determine transverse separation forces and design nozzles to better withstand operational loads. Marshall also uses LMS Test.Lab for ground vibration testing (GVT) of the new vehicles.

Preparations are underway for GVT of the complete Ares I craft to be conducted in 2011 using the dynamic test stand at Marshall. Tests will be performed on the "full stack," or the complete vehicle, including the first and second stage motors, fuel tanks, and crew capsule. Structural vibrations will be induced using up to six hydraulic or electrodynamic shakers delivering random and sine excitations. LMS Test.Lab can provide engineers with critical test data including FRFs, natural frequencies, damping values, and mode shapes to evaluate how the structure will likely vibrate during liftoff, stage separation, and subsequent phases of the flight.

The LMS SCADAS 260-channel front-end is also one of NASA's large modal data-acquisition systems. The high channel count enables the modal test measurements in fewer test runs. Measuring multiple functions simultaneously allows them to obtain FRFs as well as associated cross spectrums, auto powers, and time data in parallel instead of having to run separate tests. The modal test team plans to complete the Ares GVT in only three test sets versus up to eight runs needed for comparable tests on the Saturn and shuttle vehicles using a system with far fewer channels.

The team also makes extensive use of LMS PolyMAX software, which automatically highlights resonances and provides consistent results that could otherwise vary due to subjective interpretation. In addition, animated operational deflection shape features show how the structure may bend and twist at various frequencies so engineers have deeper insight into dynamic structural behavior.

LMS is focused on the mission critical performance attributes in key manufacturing industries, including structural integrity, system dynamics, handling, safety, reliability, comfort, and sound quality. From this work, LMS engineers gained knowledge that will help develop tomorrow's rocket propulsion systems and can also be used for engineering applications in a wide range of other industries. By providing onsite support for tests, the LMS technical support and development staff seize opportunities like the work with NASA to expand their knowledge of tests and dynamics in real-world applications.

Product Outcome

As the pool of companies and agencies testing rockets is limited, knowledge related to the execution and optimization of design resulting from such tests is likewise small. Exposure to this data will lead to better modeling and simulation, resulting in better and safer products for the public. This approach of creating system models based on modal test data is useful in research and development studies of similar structures that are difficult to model and whose dynamic behavior is of primary interest. By working with NASA, LMS engineers gained access to uncommon test data to enhance and refine their product to help companies test future processes and designs.

In one recent example of the benefit of amalgamating experiences into an integrated platform, the Spanish division of the European Aeronautic Defence and Space Company, Construcciones Aeronáuticas S.A. (EADS-CASA), also Spain's leading aeronautical company, implemented LMS Test.Lab and PolyMAX tools to accelerate its ground vibration testing. This aircraft testing process included a series of tests to detect the aircraft resonances as a verification of the aircraft safety and reliability before the first actual test flights. Overall, the LMS Test.Lab GVT solution and the successful deployment and technology transfer project allowed EADS-CASA to realize considerable savings in time and resources on the Airbus A330 Multi-Role Tanker Transport project.

LMS computer simulation and modeling expertise has also been applied to motorcycle safety and stability. Engineers at BMW Motorrad employed LMS DADS mechanical system simulation software to create virtual prototypes of vehicles and mechanical systems. While LMS DADS included a tire model, motorcycle tires can roll up to 50 degrees, creating forces not captured in the conventional model. BMW engineers used the program's open architecture to write in two subroutines to measure wobble, weave, and kickback. The first subroutine modeled throttle, brake, and handlebar inputs by a virtual rider. The second modeled tires and their interaction with the pavement, including variables for the frame; lower and upper forks; Telelever, a front suspension design unique to BMW motorcycles; front and rear wheels; rear swing arm; and other components. This model proved remarkably accurate in evaluating motorcycle design—engineers consider the simulation results at least as accurate as measurements taken on the test track, but with less invested time and expense. BMW is now able to specify structural design requirements, such as stiffness and mass distribution, which will ensure greater stability and safety of the end product. ❖

Test.Lab®, Virtual.Lab®, SCADAS®, and PolyMAX® are registered trademarks of LMS International NV.

Telelever™ is a trademark of BMW AG Motorrad.

LMS software and data-acquisition hardware help test, analyze, and optimize a vehicle's ability to withstand the vibration loads experienced during missions.

Software Performs Complex Design Analysis

Originating Technology/NASA Contribution

Designers use computational fluid dynamics (CFD) to gain greater understanding of the fluid flow phenomena involved in components being designed. They also use finite element analysis (FEA) as a tool to help gain greater understanding of the structural response of components to loads, stresses and strains, and the prediction of failure modes.

Automated CFD and FEA engineering design has centered on shape optimization, which has been hindered by two major problems: 1) inadequate shape parameterization algorithms, and 2) inadequate algorithms for CFD and FEA grid modification.

Working with software engineers at Stennis Space Center, a NASA commercial partner, Optimal Solutions Software LLC, was able to utilize its revolutionary, one-of-

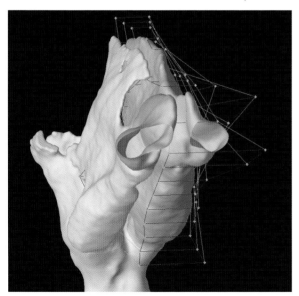

With traditional design software, reshaping an object can be a very tedious job, requiring point-by-point adjustment of hundreds or thousands of grid points. Sculptor makes this process simple and intuitive.

a-kind arbitrary shape deformation (ASD) capability—a major advancement in solving these two aforementioned problems—to optimize the shapes of complex pipe components that transport highly sensitive fluids.

The ASD technology solves the problem of inadequate shape parameterization algorithms by allowing the CFD designers to freely create their own shape parameters, therefore eliminating the restriction of only being able to use the computer-aided design (CAD) parameters.

The problem of inadequate algorithms for CFD grid modification is solved by the fact that the new software performs a smooth volumetric deformation. This eliminates the extremely costly process of having to remesh the grid for every shape change desired. The program can perform a design change in a markedly reduced amount of time, a process that would traditionally involve the designer returning to the CAD model to reshape and then remesh the shapes, something that has been known to take hours, days—even weeks or months—depending upon the size of the model.

Partnership

Optimal Solutions Software (OSS) LLC, of Provo, Utah, and Idaho Falls, Idaho, creates highly innovative engineering design improvement products to enable engineers to more reliably, creatively, and economically design new products in high-value markets.

The company entered into a **Small Business Innovation Research (SBIR)** contract with Stennis, under which it extensively used its ASD software to improve pressure loss, velocity, and flow quality in the pipes utilized by NASA. The product is available under the trade name Sculptor.

Because of the funding from the SBIR program and the technical contributions from its NASA counterparts, OSS was able to take the technological know-how and commercial successes gained from this project and effectively commence the next-phase step into the marketplace.

According to Mark Landon, OSS's president, "We thoroughly enjoyed working with the engineers and scientists at Stennis—they were technically very sound and extremely helpful in every aspect of the success of the project. Additionally, with the technical and funding assistance from this program, OSS was able to create new jobs, our revenue figures and sales have increased, and private investors are looking at us at this time to take us to the next stage on our road to commercialization."

Implementing Sculptor's ASD technology and optimizer during the Stennis SBIR Phase II test case, OSS demonstrated a steady-state optimized resistance temperature device (RTD), which produced a smaller and more symmetrical wake, resulting in a lower drag coefficient, thus a lower moment at the base of the RTD. Remarkably, because of the aerodynamic drag reduction from the shape optimization, there was a 60 percent reduction in moment at the base of the RTD probe.

NASA applications for Sculptor include the design of spacecraft shapes; aircraft shapes; propulsion devices (nozzles, combustion chambers, etc.); pumps; valves; fittings; and other components.

With the assistance from programs such as the SBIR program, the company continues to expand its product family and add features so that its customers can create breakthrough designs and realize increased efficiency gains.

Product Outcome

Sculptor can be applied to almost any fluid dynamics problem, structural analysis problem, acoustics design, electromagnetic design, or any area where the designer needs to be able to address complex design analysis. The program performs smooth volumetric deformation and can institute design changes in seconds.

OSS has sold a license for Sculptor to Eglin Air Force Base for the design of unmanned aircraft drones and miniature aircraft. The company has also sold a license for

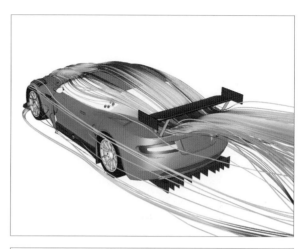

Computational fluid dynamics (CFD) is proving an effective method for testing racecars—replacing, in some instances, the traditional track and wind tunnel tests.

the software to the U.S. Navy for use with Combustion Research and Flow Technology Inc., the author of the CRUNCH CFD code, a multi-element, unstructured flow solver for viscous, real gas systems, currently in use for cavitation modeling, turbo machinery applications, and large eddy simulation.

OSS has also teamed with Engineous Software Inc., the author of the iSIGHT optimization software, to subcontract on a Phase I SBIR with Wright-Patterson Air Force Base to use Sculptor to provide the shape deformation and shape matching for fluid structure interaction solutions. Wright-Patterson has also purchased a Sculptor license for application to aircraft shape design.

Since Sculptor can be utilized in any instance where there is fluid (gas or liquid) flowing in, around, or through an object, the applications are nearly countless.

Automotive shapes and parts such as aerodynamics of the body itself, mirrors, internal flow components such as

intake manifolds, radiators, exhaust manifolds, cylinders, and air conditioning ducts all benefit from this program.

The motor sports industry is currently a big customer, with many racing teams finding a competitive edge by using Sculptor to reduce drag, improve down forces, improve engine design, and perform other design analyses, such as for brake cooling, basic flow handling, and internal combustion components.

Other sports are showing a deep interest in the Sculptor technology for such activities as golf club design, swimming and boating aerodynamics—even model airplane flight efficiencies.

Also showing great promise for using these tools are other industries such as biomedicine, which can utilize its ability to more quickly predict the effect of shape (anatomical) changes in the body's vascular and other bodily systems.

CFD shows pressure and streamlines along the body of this Aston Martin DBR9 GT series racecar.

To expand its domestic and worldwide presence, OSS has established a powerful distribution network, which covers North America, all of the European Union, Japan, South Korea, and China. ❖

Sculptor™ is a trademark of Optimal Solutions Software LLC.

Crunch CFD® is a registered trademark of Combustion Research and Flow Technology Inc.

iSIGHT™ is a trademark of Engineous Software Inc.

NASA makes our world safer. The technologies featured in this section:

- Protect deep-sea divers
- Monitor strain and reduce cost
- Protect firefighters, military, and civilians
- Ensure safe environments

Space Suit Technologies Protect Deep-Sea Divers

Originating Technology/NASA Contribution

Working on NASA missions allows engineers and scientists to hone their skills. Creating devices for the high-stress rigors of space travel pushes designers to their limits, and the results often far exceed the original concepts. The technologies developed for the extreme environment of space are often applicable here on Earth.

Some of these NASA technologies, for example, have been applied to the breathing apparatuses worn by firefighters, the fire-resistant suits worn by racecar crews, and, most recently, the deep-sea gear worn by U.S. Navy divers.

Partnership

Paragon Space Development Corporation, founded in 1993, is located in Tucson, Arizona. This firm is a woman-owned small business, specializing in aerospace engineering and technology development, and is a major supplier of environmental control and life support system and subsystem designs for the aerospace industry. Paragon has proven itself expert in thermal control for spacecraft in orbit and during reentry, as well as for hypervelocity aircraft.

In recent years, Paragon has worked on several different projects that benefit NASA and the space community. Through a NASA-funded **Small Business Innovation Research (SBIR)** contract, Paragon utilized its unique thermal analysis and structural design capabilities to develop a new, reduced-weight radiator system for use on the Orion Crew Exploration Vehicle, other next-generation spacecraft, and commercial vehicles.

Paragon credits the Arizona Department of Commerce and the Governor's FAST grant award (Federal and State Technology Partnership program) for the seed funds that led to the NASA SBIR award. The FAST grant program is funded by the U.S. Small Business Administration and is focused on capturing Federal grants for competitive small businesses in each state, creating new jobs and new markets that lead to a better and stronger economy by keeping high-technology jobs in America. Paragon used its $5,000 FAST grant award to write its initial proposal to NASA, a partnership that led to continued research grants and development opportunities.

Other developments resulting from NASA research include Paragon's Environmental Control and Life Support Human-rating Facility, which the company designed to test emerging life support system designs for suborbital and orbital spacecraft, and the solid oxide electrolysis (SOE) technology, which is under continued development as a Phase II NASA SBIR. The SOE technology directly breaks down the carbon dioxide given off by the crew of a space vehicle and produces oxygen. This is the only known technology with the potential to supply all the crew's oxygen needs directly from the crew's metabolic byproducts, significantly saving spacecraft logistical mass. Another NASA project resulted in the development of the metabolic heat temperature swing absorption, which incorporates the technology innovation of using the metabolic heat generated by a space-suited astronaut to absorb and purge carbon dioxide from the breathing loop.

"[Our] partnership with NASA is growing rapidly and has many facets, from spinoffs that protect and enable the war fighter in extreme environments, to technologies that will be used on the Orion spacecraft, and support astronauts on the Moon and Mars," explained Taber MacCallum, CEO and chairman of the board for Paragon. "NASA is the premier technology organization, and partnering with NASA is a key part of Paragon's business plan. In our experience, what you put into the partnership determines what you get out of it. Industry's partnership with NASA is a central component in maintaining America's technical preeminence and high-technology jobs."

Similarly, NASA is likely to rely on such commercial space services during the interval between the retirement

Paragon Space Development Corporation used its NASA know-how to create diving suits capable of protecting deep-sea divers from hazardous environments.

of the space shuttle and the initial flight of Orion and its Ares launch vehicle.

Product Outcome

Navy divers are called on to work in extreme and dangerous conditions. The high pressure of deep diving, toxic chemical spills, hot waters of the Persian Gulf, and chemical warfare agents make for some of the most hazardous working environments on Earth. As such, the Navy requested a diving system that will not fail when exposed to chemicals and would create an impermeable protective shell around the diver. Paragon's extensive experience providing life support in extreme environments assisted in the development of a line of such products to protect Navy divers against hazardous materials; in particular, the successful design of a diving suit that now has the potential for use in commercial diving.

In designing the suit, Paragon applied its understanding of air flow in a space suit helmet, use of an

The Navy has since requested five units for field testing prior to outfitting all Navy dive suits with the Paragon product. Other products under development at Paragon include individual or collective protection systems designed for use in land vehicles and structures.

MacCallum says, "Bringing space technology back to Earth, we provide space suit-like protection for divers working in hazardous environments ranging from chemical and biological warfare agents, to the toxic environment of a shipwreck or chemical spill. Conversely, our technology is now being considered as a way to protect municipal water supplies from being contaminated by divers servicing potable water tanks." ❖

In addition to the obvious dangers inherent in deep-sea diving, divers often encounter hazardous environments, such as waters contaminated by oil and chemical spills, pathogenic microbes, extreme temperatures, and possibly even biological or chemical warfare agents.

umbilical to support an astronaut during a spacewalk, cooling undergarment systems to remove excess body heat, computer codes for thermal and airflow analysis, and materials that have been developed for the aerospace industry that are resistant to extreme chemical and temperature environments.

According to MacCallum, the Paragon suit provides "space suit-like" isolation, delivering safe breathing air to the diver. The surface-supplied system collects exhaled air and returns it to the surface to eliminate ingress pathways of hazardous agents through the regulator. The materials, including all soft goods, are impermeable. The development unit has completed unmanned testing,

and the human-rated prototype has completed manned testing, having been evaluated by Navy divers at the Navy Experimental Diving Unit facility in Pensacola, Florida.

"The contaminated water diving technology that Paragon developed for the Navy came about as a result of our partnership with NASA. We are able to protect the war fighter and enable missions in extreme pressure, temperature, and chemical environments because NASA paved the way with space suit technologies and the operational know-how that allows astronauts to work in the extreme environment of space."

More recently, Paragon provided the Navy with a prototype of its Regulated Surface Exhaust Diving System.

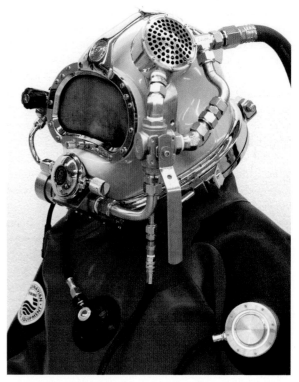

The specially designed suits provide deep-sea divers with space suit-like protection.

Fiber Optic Sensing Monitors Strain and Reduces Costs

Originating Technology/NASA Contribution

In applications where stress on a structure may vary widely and have an unknown impact on integrity, a common engineering strategy has been overbuilding to ensure a sufficiently robust design. While this may be appropriate in applications where weight concerns are not paramount, space applications demand a bare minimum of mass, given astronomical per-pound launch costs. For decades, the preferred solution was the tactic of disassembly and investigation between flights. Knowing there must be a better way, Dr. Mark Froggatt, of Langley Research Center, explored alternate means of monitoring stresses and damage to the space shuttle.

While a tear-it-apart-and-have-a-look strategy was effective, it was also a costly and time consuming process that risked further stresses through the very act of disassembly and reassembly. An alternate way of monitoring the condition of parts under the enormous stresses of space flight was needed. Froggatt and his colleagues at Langley built an early-warning device to provide detailed information about even minuscule cracks and deformations by etching a group of tiny lines, or grating, on a fiber optic cable five-thousandths of an inch thick with ultraviolet light. By then gluing the fiber to the side of a part, such as a fuel tank, and shining a laser beam down its length, reflected light indicated which gratings were under stress. Inferring this data from measurements in light rather than in bonded gauges saved additional weight. Various shuttle components now employ the ultrasonic dynamic vector stress sensor (UDVSS), allowing stress detection by measuring light beamed from a built-in mini-laser.

By measuring changes in dynamic directional stress occurring in a material or structure, and including phase-locked loop, synchronous amplifier, and contact probe, the UDVSS proved especially useful among manufacturers of aerospace and automotive structures for stress testing and design evaluation. Engineers could ensure safety in airplanes and spaceships with a narrower, not overbuilt, margin of safety. For this development, in 1997, Discover Magazine named Froggatt a winner in the "Eighth Annual Awards for Technological Innovation" from more than 4,000 entries.

Partnership

Froggatt continued his work in monitoring stresses of fiber optic components, accessories, and networks through optical monitoring at Luna Technologies, a division of Luna Innovations Incorporated, based in Blacksburg, Virginia. At Luna, he headed a team that developed the Optical Backscatter Reflectometer (OBR) with distributed sensing. The OBR is a fiber optic diagnostic tool that locates and troubleshoots splices, breaks, and connectors in fiber assemblies. In addition, it transforms standard telecom-grade fiber into a distributed strain and temperature sensor.

In 2002, Luna Innovations Incorporated entered into a licensing agreement with NASA for patent rights to products developed from Froggatt's earlier work on the UDVSS. Since that initial licensing, Luna has released the Optical Vector Analyzer (OVA), Distributed Sensing System (DSS), and the OBR platforms.

Product Outcome

Luna now has several lines of sensing and instrumentation products that are sold under the branded name of Luna Technologies. The Luna Technologies brand offers advances in optical test products helping the communications industry to increase productivity and improve component characterization while dramatically reducing the development process and production costs. Fiber optic sensing instruments includes the OVA group, a set of instruments for linear characterization of single-mode optical components, and two different techniques for distributed sensing: the DSS, which uses Fiber Bragg Gratings (FBG), and the OBR, which uses standard telecom-grade optical fiber.

First profiled in *Spinoff* 2002, the OVA is the first instrument on the market that is capable of full and complete all-parameter linear characterization of single-mode optical components. The OVA further evolved into a fast, accurate, and economical suite of tools for loss, dispersion, and polarization measurement of modern optical networking equipment, including FBG, arrayed waveguide gratings, free-space filters, tunable devices, amplifiers, couplers, and specialty fiber.

The DSS is a fiber optic sensing tool for taking distributed measurements of temperature and strain. The DSS uses swept-wavelength interferometry to simultaneously interrogate thousands of sensors integrated in a single fiber. These sensors consist of discrete FBG point sensors which can each reflect the same nominal wavelength. As such, the sensors can be fabricated on the draw tower, eliminating the need for individual grating fabrication. DSS applications include structural monitoring for naval, aerospace, and civil structures; temperature profile monitoring in extreme environments; pipeline shift and leak detection; and electrical power line sag and temperature monitoring.

The OBR offers unprecedented diagnostic capabilities and is a true high-resolution optical time domain reflectometer designed specifically for qualifying fiber components, modules, and cable assemblies for telecommunications, avionics/military-aerospace, and fiber-sensing applications. Through distributed sensing, the OBR can transform standard telecom-grade fiber into a high-spatial-resolution strain and temperature sensor. Using swept wavelength interferometry (SWI) to measure the Rayleigh backscatter as a function of length in optical fiber with high-spatial resolution, the OBR measures shifts and scales them to give a distributed temperature or strain measurement. The SWI approach enables robust and practical distributed temperature and strain measurements in standard fiber with millimeter-scale spatial resolution over hundreds of meters of fiber

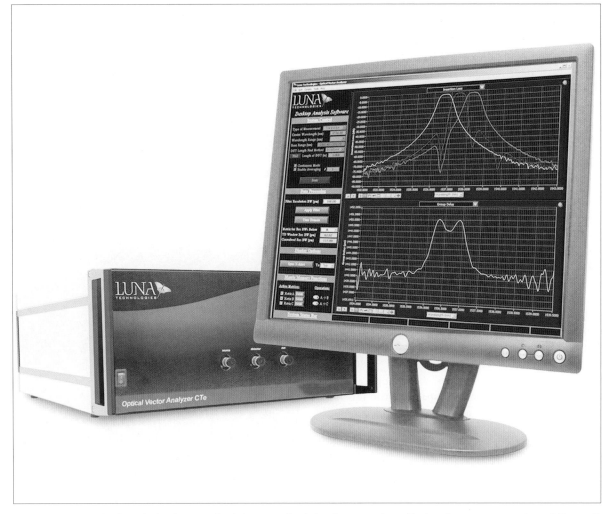

The Optical Vector Analyzer is the first completely integrated solution for measuring critical performance parameters of fiber optic components and modules.

with strain and temperature resolution as fine as 1 µstrain and 0.1 °C.

As with the other fiber optic monitoring tools, OBR provides isolation of faults and problems well before final testing, saving hours in rework and expenses in yield loss.

These abilities netted the OBR some prestigious awards:

- 2005 Lightwave "Attendees' Choice Award" in the Test Equipment category for the second consecutive year.

- 2005 Frost & Sullivan "Optical Product of the Year Award," as the industry's most sensitive frequency domain reflectometer.

- 2007 "R&D 100" award from the editors of R&D Magazine as one of the 100 most technologically significant new products introduced into the marketplace in the last year. Past "R&D 100" awards acknowledgements have included the automated teller machine (ATM), the fax machine, the NicoDerm antismoking patch, and high-definition television (HDTV).

"The 'R&D 100' award provides a mark of excellence known to industry, government, and academia as proof that a product is one of the most innovative of the year across a broad range of technologies," said Brian Soller, president of the Products Division at Luna. "This is the first year Luna has submitted an award nomination to the R&D 100, and we are honored to have our test and measurement instrument selected as part of this truly elite group."

Released in March 2007, the OBR 4400 is an upgraded version of the OBR instrument, with enhanced capabilities in a more compact design. Range has been increased to 2 kilometers, still with millimeters of resolution, and users can monitor the effects from component-level heating in optical amplifiers to strain and load redistribution in aircraft harnesses. Other applications include temperature monitoring inside telecommunications cabinets and enclosures, and a feature that allows users to identify the location in fiber assemblies simply by touching the fiber. With a small, easily transportable platform, the OBR 4400 provides the user with precision reflectometry and unprecedented optical-module inspection and diagnostic capabilities. Luna Technologies also recently introduced a tunable laser, a precision reflectometer, and an optical switch to round out their product offering. ❖

Optical Vector Analyzer™, Optical Backscatter Reflectometer™, and Distributed Sensing System™ are trademarks of Luna Technologies. NicoDerm® is a registered trademark of GlaxoSmithKline.

Polymer Fabric Protects Firefighters, Military, and Civilians

Originating Technology/NASA Contribution

Insulating and protecting astronauts from temperature extremes, from the 3 K (-455 °F) of deep space to the 1,533 K (2,300 °F) of atmospheric reentry, is central to NASA's human space flight program. While the space shuttle and capsule vehicles necessarily receive a great deal of thermal barrier and insulation protection, at least as much attention is also paid to astronaut clothing and personal gear. NASA has spent a great deal of effort developing and refining fire-resistant materials for use in vehicles, flight suits, and other applications demanding extreme thermal tolerances, and kept a close eye on the cutting edge of high-temperature stable polymers for its entire 50-year history.

In the late 1950s, Dr. Carl Marvel first synthesized Polybenzimidazole (PBI) while studying the creation of high-temperature stable polymers for the U.S. Air Force. In 1961, PBI was further developed by Marvel and Dr. Herward Vogel, correctly anticipating that the polymers would have exceptional thermal and oxidative stability. In 1963, NASA and the Air Force Materials Laboratory sponsored considerable work with PBI for aerospace and defense applications as a non-flammable and thermally stable textile fiber.

On January 27, 1967, the severity and immediacy of the danger of fire faced by astronauts was made terribly clear when a flash fire occurred in command module 012 during a launch pad test of the Apollo/Saturn space vehicle being prepared for the first piloted flight, the AS-204 mission (also known as Apollo 1). Three astronauts, Lieutenant Colonel Virgil I. Grissom, a veteran of Mercury and Gemini missions; Lieutenant Colonel Edward H. White II, the astronaut who had performed the first U.S. extravehicular activity during the Gemini program; and Lieutenant Commander Roger B. Chaffee, an astronaut preparing for his first space flight, died in this tragic accident.

A final report on the tragedy, completed in April 1967, made specific recommendations for major design and engineering modifications, including severely restricting and controlling the amount and location of combustible materials in the command module and the astronaut flight suits. NASA intensified its focus on advanced fire-resistant materials, and given the Agency's existing familiarity with the fabric and its inventor, one of the first alternatives considered was PBI.

Partnership

NASA contracted with Celanese Corporation, of New York, to develop a line of PBI textiles for use in space suits and vehicles. Celanese engineers developed heat- and flame-resistant PBI fabric based on the fiber for high-temperature applications. The fibers formed from the PBI polymer exhibited a number of highly desirable characteristics, such as inflammability, no melting point, and retention of both strength and flexibility after exposure to flame. The stiff fibers also maintained their integrity when exposed to high heat and were mildew, abrasion, and chemical resistant.

Throughout the 1970s and into the 1980s, PBI was instrumental to space flight, seeing application on Apollo, Skylab, and numerous space shuttle missions. Applications ran the gamut from the intended applications in astronaut flight suits and clothing, to webbing, tethers, and other gear that demanded durability and extreme thermal tolerance.

Product Outcome

In 1978, PBI was introduced to fire service in the United States, and Project FIRES (Firefighters Integrated Response Equipment System) lauded a recently developed outer shell material for turnout gear, PBI Gold. In 1983, PBI fibers were made commercially available and a dedicated production plant opened in Rock Hill, South Carolina, to meet demand. In 1986, NASA *Spinoff* chronicled this first phase of PBI's history, and

The PBI plant, located in Rock Hill, South Carolina, produced its first commercial bale of PBI fiber on March 18, 1983.

Marvel was awarded the "National Medal of Science" by President Ronald Reagan.

Since 1986, PBI has undergone a steady evolution into countless military and civilian applications and established a distinct profile and reputation in the fire retardant materials industry. In 2005, Celanese Corporation sold the PBI fiber and polymer business to PBI Performance Products Inc., of Charlotte, North Carolina, which is under the ownership of the InterTech Group, of North Charleston, South Carolina.

Produced by a dedicated manufacturer that takes great pride in the history and future of the product, the fabrics incorporating PBI have become prominent players in such diverse applications as firefighting and emergency response, motor sports, military, industry, and (still) aerospace. PBI Performance Products now offers two distinct lines: PBI, the original heat and flame resistant fiber; and Celazole, a family of high-temperature PBI polymers available in true polymer form.

- PBI fabric withstands the dangers associated with firefighting, arc flash, and flash fire. In 1992, lightweight PBI fabrics were adapted for flame-resistant work wear for electric utility and petrochemical applications, and are now providing flame protection for U.S. Army troops in Afghanistan and Iraq. Short-cut PBI fibers were introduced for use in automotive braking systems and PBI staple fibers are employed as fire blocking layers in aircraft seats.

- PBI Gold blends 40 percent thermal-resistant PBI fibers with 60 percent high-strength aramid, resulting in a fabric which does not shrink, become brittle, or break open under extreme heat and flame exposure. PBI Gold provides firefighters and industrial workers with superior protection and meets or exceeds every National Fire Protection Association (NFPA) and EN 469 (rating standard for protective clothing for firefighters) requirement. In 1994, the New York City Fire Department specified the use of PBI Gold fabric engineered in black for their turnout gear. Over the last 10 years, PBI Gold has grown internationally, with major industrial, military, and municipal fire brigades specifying the product across Europe, the Middle East, Asia, Australia, and the South Pacific.

- PBI Matrix employs a "power grid," a durable matrix of high-strength aramid filaments woven into the PBI Gold fabric to enhance and reinforce its resistance

Andre Baur, a firefighter instructor in Switzerland, runs out of a training fire that has gotten out of hand. Like many other "Golden Knights" around the world, Andre escaped with only minor injuries.

to wear and tear while retaining its superior flame and heat protection. In 2003, PBI Matrix was commercialized and introduced in the United States as the next-generation PBI for firefighter turnout gear. In 2008, Matrix will be introduced in Europe.

- PBI TriGuard fabric is a three-fiber blend of PBI, Lenzing FR, and MicroTwaron designed for flame protection, comfort, and durability. This advanced fabric meets or exceeds all U.S. Department of Labor Occupational Safety and Health Administration (OSHA) and NFPA standards and is certified for wildlands, special operations, and motorsports applications, as well as the petrochemical, gas utility, and electric utility industries. PBI TriGuard and PBI Gold knits are now in use at several major motorsport racetracks around the country.

- Celazole T-Series is a form-, shape-, and an injection-moldable blend of PBI and PEEK (polyetheretherketone) polymers.

- Celazole U-Series utilizes PBI's high-heat dimensional stability, strength, and chemical resistance to allow

it to be formed into parts and used in the tools that produce flat panel displays and in the plasma etch chambers used to make semiconductor wafers.

New applications for PBI are continuing to come to light in new fields that demand material stability at high temperatures. PBI is now being developed into high-temperature separation membranes that increase efficiency in ethanol production and separate carbon dioxide from natural gas for carbon dioxide sequestration, and will see application in hydrogen fuel cells. PBI in short-cut form has also been used as a safe and effective replacement for asbestos. Fittingly, PBI may also return to space as part of NASA's Constellation Program, as the polymer once applied for space suits in the Apollo and Skylab missions is under consideration for use as insulation material in the rocket motors for NASA's next generation of spacecraft, the Ares I and Ares V rockets. ❖

PBI TriGuard™ is a trademark, and PBI Gold®, PBI Matrix®, and Celazole® are registered trademarks of PBI Performance Products Inc. Lenzing FR® is a registered trademark of Lenzing Fibers GmbH. MicroTwaron™ is a trademark of Akzo N.V.

Advanced X-Ray Sources Ensure Safe Environments

Originating Technology/NASA Contribution

Successfully sustaining life in space requires closely monitoring the environment to ensure the health of the crew. Astronauts can be more sensitive to air pollutants because of the closed environment, and pollutants are magnified in space exploration because the astronauts' exposure is continuous. Sources of physical, chemical, and microbiological contaminants include humans and other organisms, food, cabin surface materials, and experiment devices.

One hazard is the off-gassing of vapors from plastics and other inorganic materials aboard the vehicle, vividly illustrated by Skylab—in 1973, NASA scientists identified 107 volatile organic compounds in the air inside the Skylab space station. All synthetic materials exude low-level gasses, known as off-gassing; when these chemicals are trapped in a closed environment, as was the case with the Skylab, the inhabitants may become ill. To avoid this, air sampling systems on the International Space Station (ISS) periodically check the air for potential hazards. Advanced, high-efficiency particulate air filters and periodic filter cleanings have been successful in keeping harmful vapors out of the air. Other significant contaminants that pose hazards to the crew are microbial growth, both bacterial and fungal; air, water, and surface sampling by the crew in conjunction with periodic cleaning keep the microbial levels on the ISS in check.

To monitor microbial levels, crew members use devices called grab sample containers, dual absorbent tubes, and swabs to collect station air, water, and surface samples and send them to Earth for detailed analysis and identification every 6 months. This data provides controllers on Earth detailed information about the type of microbial contaminants on board the ISS. The controllers can then give additional direction to the crew on sanitation if increased microbial growth is identified.

Missions to the Moon and Mars will increase the length of time that astronauts live and work in closed environments. To complete future long-duration missions, the crews must remain healthy in these closed environments; hence, future spacecraft must provide even more advanced sensors to monitor environmental health and accurately determine and control the physical, chemical, and biological environment of the crew living areas and their environmental control systems.

Partnership

Ames Research Center awarded inXitu Inc. (formerly Microwave Power Technology), of Mountain View, California, a **Small Business Innovation Research (SBIR)** contract to develop a new design of electron optics for forming and focusing electron beams that is applicable to a broad class of vacuum electron devices.

This project resulted in a compact and rugged X-ray tube with a carbon nanotube (CNT) cold cathode with a circular electron beam that is focused to a diameter of less than 80 microns. The performance, durability, and operating life of CNT cathodes was enhanced by inXitu working in cooperation with Ames; Oxford Instruments, of Scotts Valley, California; and Xintek Inc., of Research Triangle Park, North Carolina; among others. inXitu constructed an automated system for screening up to 10 CNT cathodes at once. Performance data from these tests helped CNT cathode researchers and developers improve tolerance to device processing, uniformity, and stability of performance within a given lot, enhancing performance of electron beam sources and ionizers in addition to other classes of X-ray tubes. This technology provides:

- Inherently rugged and more efficient X-ray sources for material analysis

- A miniature and rugged X-ray source for smaller rovers on future missions

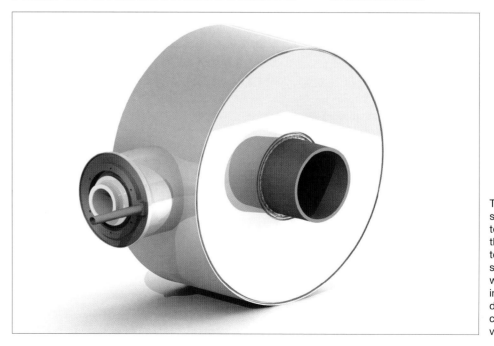

The electron beam source can be scaled to fit any duct size and the flanges adapted to mate with existing systems. Reactions with the electron beam in the duct section destroy or neutralize contaminants entering via the airstream.

- Compact electron beam sources to reduce undesirable emissions from small, widely distributed pollution sources and remediation of polluted sites
- Large area emitters for new X-ray sources in future baggage scanning systems

Researchers derived a mathematical distribution function for the beam with a purpose-built electron beam analyzer, which characterized the unique behavior of electron beams emitted from CNT cathodes. A boundary element computer incorporated the distribution function code to design the electron optics, with an electrostatically focused electron gun and magnetic lens to focus the electron. The final X-ray tube consists of rugged metal ceramic construction welded into a 2-inch-diameter package along with a 40 kV power supply. This design forms a hermetic package that can withstand severe environmental stresses encountered during launch, landing, and operation in space.

NASA will apply this technology in versatile X-ray instruments capable of operating in both a fluorescence or diffraction mode for in situ analysis of rocks and soils of the solid bodies in the solar system to determine their atomic constituents and mineralogy. Other applications of this technology include purifying air in space and Moon base stations, eliminating toxic products and biological toxins in aircraft, enhancing chemical reactions in space-based manufacturing, and sterilization of material to be returned to Earth or taken to space from Earth. inXitu was awarded a Phase III SBIR contract in 2006 to continue this work.

Product Outcome

Oxford's X-ray Technology Group provides laboratory space and production support for continuing development and commercialization of advanced CNT-based vacuum sources. The company produced Eclipse 1 and Eclipse 2 X-ray sources from inXitu's prototype that was used in hand-held and portable fluorescence spectrometers

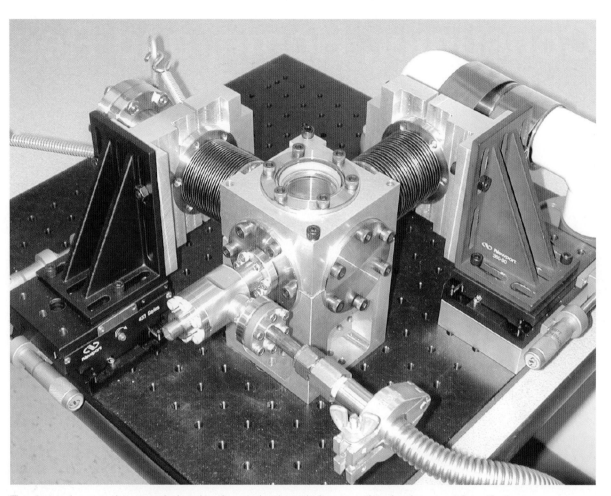

The electron beam analyzer was designed to characterize the emission properties of carbon nanotube cathodes.

for in situ analysis of materials and surfaces. The Eclipse 2 X-ray tube was applied in equipment for monitoring paper coating and other high-speed processes.

Next-generation baggage and cargo screening systems employ CNT cold cathode X-ray sources, promising increased throughput, reduced false alarm rates, reduced power consumption, reduced heat load, reduced size and weight, and improved ruggedness and responsiveness over existing thermionic X-ray sources. Additional commercial applications include air purification; odor elimination; non-burning destruction of evaporated hydrocarbons from fuel tanks and painting operations; soil and groundwater remediation; flue gas cleaning; and chemical reaction enhancements, such as increasing fuel efficiency and reducing ink drying speed, as well as surface sterilization. ❖

Consumer, Home, and Recreation

NASA research improves our quality of life. The technologies featured in this section:

- Equip boat owners
- Make panoramic photography a snap
- Advance gardening at home and in space
- Reduce drag, break records
- Render 360° views
- Capture our imagination
- Maximize power and dependability
- Fortify nutrition worldwide
- Insulate missions and consumer products

Wireless Fluid-Level Measurement System Equips Boat Owners

Originating Technology/NASA Contribution

While developing a measurement acquisition system to be used to retrofit aging aircraft with vehicle health monitoring capabilities, Langley Research Center's Dr. Stanley E. Woodard and Bryant D. Taylor, of ATK Space, developed a novel wireless fluid-level measurement system.

Current fluid-level measurement methods, which see widespread application, have significant drawbacks, including limited applicability of any one fluid-level sensor design; necessity for each sensor to be supplied power via a direct electrical connection and have a physical connection to extract a measurement; and need for a data channel and signal conditioning electronics dedicated to each sensor. Use of typical wired systems results in other shortcomings, such as logistics for adding or replacing sensors, weight, and the potential for electrical arcing and wire degradations.

The wireless fluid-level measurement system that Woodard and Taylor developed, however, uses sensors that are simple: passive inductor-capacitor circuits. The system is laid out in seven U.S. and international patents and patents-pending, collectively resulting in an inexpensive and safe wireless fuel measurement system that can be used to measure the volume of any fluid at any orientation. A key safety feature unique to the system is that it allows the sensors to be completely encapsulated so that the fuel level can be measured with neither the fuel nor fuel vapors coming in contact with any electrical components of the system, thus eliminating the potential for combustible fuel vapors being ignited by arcing from damaged electrical components.

Woodard explains, "This technology eliminates many of the causes of the TWA Flight 800 and Swissair Flight 111 accidents. These accidents resulted in the loss of 459 lives. In both cases, damage to a direct electrical line from the aircraft power system to a fuel probe inside a fuel tank

containing combustible fuel vapors was a critical link in a chain of events that led to these tragedies."

The sensor is also not subject to the mechanical failure possible when float and lever-arm systems are used—system sensors are powered by oscillating magnetic fields; once electrically excited, each sensor produces its own magnetic field response, the frequency of which corresponds to the amount of fluid within the sensor's electric field. The system can be used to measure any fluid in any container, including those on aircraft, cars, boats,

trains, trucks, or even the space shuttle and satellites. In addition to the safety features, the system is able to automatically recalibrate for new fuels, a feature that is becoming more attractive as the number of flex-fuel vehicles increases.

One especially key advantage of this technology is that it can be used with any system requiring fluid-level measurement, regardless of the fluid being measured. The sensor design can be modified for measuring the level of any fluid or non-gaseous fluid substance that can

In addition to giving boat operators a highly accurate reading of fuel levels, this sensor detects water in the bottom of the tank, whether it is mixed with ethanol, gasoline, diesel, or two-stroke oil.

be stored in a nonconductive reservoir. The inventors demonstrated this by measuring levels of ammonia, liquid nitrogen, salt water, tap water, transmission fluid, bleach, sugar water, and hydrochloric acid—all elements that would easily destroy most electronics. The system's ability to take accurate measurements of the level of non-liquids has been tested with powdered sugar and ground corn. Perhaps most importantly, it has been used to measure the levels of a variety of petroleum products, which led to its first commercial application.

Partnership

The NASA technology was of interest to Tidewater Sensors LLC, of Newport News, Virginia, because of its many advantages over conventional fuel management systems, including its ability to provide an accurate measurement of volume while the boat is experiencing any rocking motion due to waves or people moving about

on the boat. Like a conventional float gauge, it is quick and easy to install, but unlike the float gauge, this device has no moving parts, is sealed from the elements, and allows the boat owner to use any size or shape tank and still get an accurate reading. The system also introduces no electronics into the tank and has no connections at the sensor that need grounding. These advantages led the company to license the novel fluid-level measurement system from NASA for marine applications.

Product Outcome

The Tidewater Sensors commercial version of the NASA measurement system is available under the name TS1500. The non-moving probe contains both the antenna and the sensor as a single unit that is easily interfaced to any of the standard fuel display gauges used. The TS1500 is a simple, safe, easy-to-operate tool that prevents expensive motor damage and helps prevent boaters from getting stranded due to motor failure.

If the TS1500 detects water, it alerts the operator with both an audible and visual alarm: the machine beeps and the fuel gauge fluctuates rapidly between empty and full. Unlike other water sensors, which require that the water be mixed with the fuel and the boat be run for a few minutes before they will operate, this sensor will alert the operator before he leaves the dock. This means that boat operators can avoid an engine-stopping combination of water and fuel in open waters. The TS1500 sounds the alarm as soon as the engine is turned on, if water is present, or as soon as the sensor touches the water.

The product boasts several other advantages over traditional float systems or capacitor sensors: it is highly accurate; senses water in gas, oil, or diesel; and uses a specially formulated rubber gasket capable of withstanding ethanol, as opposed to typical methods that provide rough measurements of fluid in the tank and use cork or butyl rubber stoppers that corrode when exposed to ethanol, leading to leaks. The TS1500 also provides linear measurements of tank capacity, as opposed to the swinging arm of a non-linear measurement, which leaves the needle indicating full for a longer period of time than may be accurate and then moves quickly toward the "E."

Tidewater Sensors has already built and field tested prototype sensors. Testing took place on boats ranging from 20 to 32 feet long, operating on coastal waters between Delaware and North Carolina. For some high-profile testing, the sensors were installed on the Hampton City (home to Langley) Fire Division's 30-foot boat, which patrols all 64 miles of the Hampton area shoreline, and the Hampton City Police boat. Sensors were also installed on a Donzi ZF powerboat owned by John Isley of the nationally syndicated morning radio show, John Boy and Billy's "The Big Show." With this high-profile testing, and boasting so many clear benefits, the NASA technology is sure to find wide commercial acceptance. ❖

Mars Cameras Make Panoramic Photography a Snap

Originating Technology/NASA Contribution

If you wish to explore a Martian landscape without leaving your armchair, a few simple clicks around the NASA Web site will lead you to panoramic photographs taken from the Mars Exploration Rovers, Spirit and Opportunity. Many of the technologies that enable this spectacular Mars photography have also inspired advancements in photography here on Earth, including the panoramic camera (Pancam) and its housing assembly, designed by the Jet Propulsion Laboratory and Cornell University for the Mars missions. Mounted atop each rover, the Pancam mast assembly (PMA) can tilt a full 180 degrees and swivel 360 degrees, allowing for a complete, highly detailed view of the Martian landscape.

The rover Pancams take small, 1 megapixel (1 million pixel) digital photographs, which are stitched together into large panoramas that sometimes measure 4 by 24 megapixels. The Pancam software performs some image correction and stitching after the photographs are transmitted back to Earth. Different lens filters and a spectrometer also assist scientists in their analyses of infrared radiation from the objects in the photographs. These photographs from Mars spurred developers to begin thinking in terms of larger and higher quality images: super-sized digital pictures, or gigapixels, which are images composed of 1 billion or more pixels.

Gigapixel images are more than 200 times the size captured by today's standard 4 megapixel digital camera. Although originally created for the Mars missions, the detail provided by these large photographs allows for many purposes, not all of which are limited to extraterrestrial photography.

Partnership

The technology behind the Mars rover PMAs inspired Randy Sargent at Ames Research Center and Illah Nourbakhsh at Carnegie Mellon University (CMU) to look at ways consumers might be able to use similar technology for more "down-to-Earth" photography and virtual exploration.

In 2005, Sargent and Nourbakhsh created the Global Connection Project, a collaboration of scientists from CMU, Google Inc., and the National Geographic Society, whose vision is to encourage better understanding of the Earth's cultures through images. This vision inspired the development of their Gigapan products.

After seeing what the Pancams and PMAs could do, Sargent created a prototype for a consumer-version of a robotic camera platform. He worked with Rich LeGrand of Charmed Labs LLC, in Austin, Texas, to design and manufacture the Gigapan robotic platform for standard digital cameras.

Product Outcome

The Gigapan robotic platform is, in essence, an intelligent tripod that enables an amateur photographer to set up detailed shots with ease. A user sets the upper-left and lower-right corners of the panorama, and the Gigapan simply will capture as many images as the user or scene requires. With this level of automation, a 500-picture panorama is no more complicated than a 4-picture panorama; only the camera's memory limits the size of the panorama.

The Global Connection Project also created two other Gigapan products: a Gigapan Web site and panorama stitching software born from the Ames Vision Workbench, an image processing and computer vision library developed by the Autonomous Systems and Robotics Area in the Intelligent Systems Division.

The robotic platform works with the stitching software by precisely manipulating and aligning each shot ahead of time. The Gigapan software complements the robotic platform by arranging the parts of the panorama (potentially hundreds of individual photographs) into a grid where they are stitched together into a single, very large Gigapan image.

The Gigapan robotic platform now enables photographers on Earth to capture and create super-sized digital panoramas.

The Global Connection Project won a 2006 "Economic Development Award" from the Tech Museum Awards for its work in creating photographic overlays for Google Earth of areas affected by natural disasters. Government workers and concerned citizens used the images on Google Earth to see which areas needed help in the aftermath of Hurricane Katrina, Hurricane Rita, and the 2005 earthquake in Kashmir.

On the Gigapan Web site, a user can display a wide bird's eye panorama and can then zoom in with impressive bug's eye high-quality detail. On first impression, a panoramic photograph on Gigapan's site might seem to be simply a wide-angle cityscape of a temple in Kathmandu. With each successive click, however, the user can zoom deeper and deeper into the photo, revealing more and more clear details: a monk hanging prayer flags on the roof of the temple and the Tibet Kitchen Restaurant and Bar a few blocks behind the temple, with a sign extolling passersby to taste their gourmet food.

As part of a continuing effort to connect people and cultures, the Global Connection Project encourages all users to upload their own panoramas from around the world on the Gigapan site. Users can explore such varied landscapes as a temple in Nepal, the Burning Man festival in the Nevada desert, a market in Guatemala, or the Boston skyline from the Charles River. Because of the much greater number of pixels, the resolution is unprecedented; the Gigapan software and robotic platforms can theoretically produce prints on 40-foot-wide paper without any loss in quality.

Whether or not photographers use the Gigapan mounts and software, anyone can upload their panoramas to the Gigapan Web site. Many users of Gigapan have uploaded standard panorama photographs, as well (although the site suggests photographs be at least 50 megabytes). This is just fine with the Gigapan and the Global Connection

Project coordinators, whose aim is simply to encourage exploration and understanding of the various cultures in our world.

The Fine Family Foundation is sponsoring work with the Global Connection Project to enable botanists, geologists, archeologists, and other scientists around the world to document different aspects of the Earth's cultures and ecosystems using Gigapan technology. Scientists are using Gigapan to document life in the upper redwood forest canopy in California, volcanoes in Hawaii, and glaciers in Norway.

There are also educational uses for the Gigapan: The Pennsylvania Board of Tourism uses Gigapan for Web site

visitors wanting to explore Civil War sites virtually. Also, in collaboration with the United Nations Educational, Scientific and Cultural Organization (UNESCO), the Global Connection Project has distributed Gigapan to students in Pittsburgh, South Africa, and the Republic of Trinidad and Tobago, encouraging them to photograph their local culture and share those panoramas with the world. "The hope is that students will be able to have deeper connections to other cultures," said Sargent.

A time-lapse Gigapan robotic mount is now in development, and a professional unit for larger SLR-style cameras may be released before the end of 2008. ❖

Gigapan™ is a trademark of Carnegie Mellon University.

Gigapan panoramic image courtesy of Jessee Mayfield.

Gigapan allows a photographer to capture extremely high-resolution panoramas, which a user can explore in depth. In this wide view of Boudhanath Stupa in Kathmandu, Nepal, it is possible to zoom all the way into the smallest, barely visible points in the picture, such as the monk standing on the roof of the temple or the sign above the Tibet Kitchen Restaurant and Bar.

Experiments Advance Gardening at Home and in Space

Originating Technology/NASA Contribution

Aeroponics, the process of growing plants suspended in air without soil or media, provides clean, efficient, and rapid food production. Crops can be planted and harvested year-round without interruption, and without contamination from soil, pesticides, and residue. Aeroponic systems also reduce water usage by 98 percent, fertilizer usage by 60 percent, and eliminate pesticide usage altogether. Plants grown in aeroponic systems have been shown to absorb more minerals and vitamins, making the plants healthier and potentially more nutritious.

The suspended system also has other advantages. Since the growing environment can be kept clean and sterile, the chances of spreading plant diseases and infections commonly found in soil and other growing media are greatly reduced. Also, seedlings do not stretch or wilt while their roots are forming, and once the roots are developed, the plants can be easily moved into any type of growing media without the risk of transplant shock. Lastly, plants tend to grow faster in a regulated aeroponic environment, and the subsequent ease of transplant to a natural medium means a higher annual crop yield. For example, tomatoes are traditionally started in pots and transplanted to the ground at least 28 days later; growers using an aeroponic system can transplant them just 10 days after starting the plants in the growing chamber. This accelerated cycle produces six tomato crops per year, rather than the traditional one to two crop cycles.

These benefits, along with the great reduction in weight by eliminating soil and much of the water required for plant growth, illustrate why this technique has found such enthusiastic support from NASA. Successful long-term missions into deep space will require crews to grow some of their own food during flight. Aeroponic crops are also a potential source of fresh oxygen and clean drinking water, and every ounce of food produced and water

Tomato plants growing in the Plant Development Habitat (PDHab) at BioServe Space Technologies. For a recent experiment on the International Space Station, the PDHab was loaded into a Commercial Generic BioProcessing Apparatus, a temperature-controlled microgravity research platform that has hosted a variety of experiments on numerous space shuttle, Mir, and ISS missions.

conserved aboard a spacecraft reduces payload weight, decreasing launch costs and freeing room for other cargo.

Partnership

In 1997, NASA teamed with AgriHouse Inc., of Berthoud, Colorado, to develop an aeroponic experiment for use on the Mir space station. Richard Stoner II, founder and president of AgriHouse, had worked with aeroponics since the late 1980s, and developed and patented a method for aeroponic crop production. AgriHouse utilized the research direction of BioServe Space Technologies, a nonprofit, NASA-sponsored Research Partnership Center located at the University of Colorado in Boulder, to assist its efforts in developing its aeroponic technology for space flight (*Spinoff* 2006). BioServe has extensive experience in space flight, having flown payload experiments on 27 shuttle missions, 2 Mir

missions (one being the above-mentioned), and several missions on the International Space Station (ISS).

To continue NASA's development of aeroponic technologies and offer a unique educational experience to students around the world, an experiment designed and built by BioServe recently flew to the ISS aboard the NASA Space Shuttle Endeavour on STS-118, in August 2007. This experiment, designed by Heike Winter-Sederroff, assistant professor of Plant Gravitational Genomics at North Carolina State University, will advance the science of growing food during long-term space expeditions and further the development of heartier varieties of tomato plants for farmers and gardeners on Earth. The experiment is also part of an educational effort involving as many as 15,000 K-12 students and teachers around the world, who will compare the growth and development of tomato plants in space with similar experiments being conducted in their own classrooms.

Essential to the success of this research was ensuring the seeds were protected on the way to the ISS, and at the same time, unable to germinate before the start of the experiment. BioServe identified an ideal medium for this transport while meeting with representatives from AeroGrow International Inc., also of Boulder, Colorado. AeroGrow's proprietary Seed Pod technology, developed for use in its AeroGarden kitchen gardening appliance, was admirably suited to the task in that it encased the seeds in a plastic framework, and thereby protected them during transit and ensured germination would not take place until proscribed by the experiment.

"AeroGrow is proud that the technologies that make our garden so simple and easy to use are being tested

The AeroGrow Seed Pod houses and protects seeds before germination and provides a platform on which to grow after germination.

for growing fresh food in space as well," said Michael Bissonnette, founder and chairman of AeroGrow. "We're thrilled to contribute to the education of so many students, and are looking forward to introducing the AeroGarden in classrooms and educational environments around the world."

Product Outcome

The use of AeroGrow Seed Pods on the ISS can be seen as the fitting fruition of an idea that sprouted several years ago. Bissonnette and colleague John Thompson were inspired by NASA experiments using aeroponic gardening to grow lettuce. The experiments reinforced that plants grown aeroponically did so significantly faster than those grown by any other method. Bissonnette and his team started working to capture this technology in a clean, simple, quick, and dependable appliance that would work in homes.

More than up to the task, AeroGrow's scientific board boasts a deep background in horticulture and aeroponics, and a depth of understanding that has helped the AeroGarden achieve such great success. For instance, Dr. Henry A. Robitaille holds undergraduate, master's, and doctorate degrees in horticulture from the University of Maryland and Michigan State University. Notably, he helped design and implement hydroponic growing systems in The Land Pavilion at Epcot Center in the Walt Disney World Resort, collaborating extensively with the NASA Controlled Ecological Life Support System research team at Kennedy Space Center.

Adapting a process as complicated as aeroponics to an automatic home appliance proved a considerable challenge and yielded impressive results. The more than 15 resulting patent applications include specialized lighting systems, nutrient tablets that nourish the plants and ensure standard pH levels regardless of municipal water supply, and the Plug & Grow Seed Pods that recently found their way to the ISS. The appeal of the AeroGarden

The AeroGarden has proven a very popular way to bring the fun and reward of growing herbs, vegetables, and flowers indoors.

has been proven in recent years, as the company has shipped over 350,000 gardens.

To this success, Bissonnette reflected, "We have succeeded in every retail channel of distribution we've rolled into, including independent culinary stores, national department store chains, independent lawn and garden and hardware chains, and have just concluded successful tests with multiple big-box retailers." Though still largely rooted in Internet and infomercial sales, AeroGardens are now found in more than 4,300 storefronts. AeroGrow has set its sights on international markets while continuing to refine and enlarge its product line. Now applied in homes and schools nationwide, and with its Seed Pods seeing application on the ISS, the fruits of NASA's work in past decades are made available in the simplicity of a kitchen countertop gardening appliance. ❖

AeroGarden™, Seed Pod™, and Plug & Grow™ are trademarks of AeroGrow International Inc.

Space Age Swimsuit Reduces Drag, Breaks Records

Originating Technology/NASA Contribution

A space shuttle and a competitive swimmer have a lot more in common than people might realize: Among other forces, both have to contend with the slowing influence of drag. NASA's Aeronautics Research Mission Directorate focuses primarily on improving flight efficiency and generally on fluid dynamics, especially the forces of pressure and viscous drag, which are the same for bodies moving through air as for bodies moving through water. Viscous drag is the force of friction that slows down a moving object through a substance, like air or water.

NASA uses wind tunnels for fluid dynamics research, studying the forces of friction in gasses and liquids. Pressure forces, according to Langley Research Center's Stephen Wilkinson, "dictate the optimal shape and performance of an airplane or other aero/hydro-dynamic body." In both high-speed flight and swimming, says Wilkinson, a thin boundary layer of reduced velocity fluid surrounds the moving body; this layer is about 2 centimeters thick for a swimmer.

Partnership

In spite of some initial skepticism, Los Angeles-based SpeedoUSA asked NASA to help design a swimsuit with reduced drag, shortly after the 2004 Olympics. According to Stuart Isaac, senior vice president of Team Sales and Sports Marketing, "People would look at us and say 'this isn't rocket science' and we began to think, 'well, actually, maybe it is.'" While most people would not associate space travel with swimwear, rocket science is exactly what SpeedoUSA decided to try. The manufacturer sought a partnership with NASA because of the Agency's expertise in the field of fluid dynamics and in the area of combating drag.

A 2004 computational fluid dynamics study conducted by Speedo's Aqualab research and development unit determined that the viscous drag on a swimmer is about

NASA helped Speedo reduce viscous drag in the new LZR Racer by performing surface drag testing and applying expertise in the area of fluid dynamics.

25 percent of the total retarding force. In competitive swimming, where every hundredth of a second counts, the best possible reduction in drag is crucially important. Researchers began flat plate testing of fabrics, using a small wind tunnel developed for earlier research on low-speed viscous drag reduction, and Wilkinson collaborated over the next few years with Speedo's Aqualab to design what Speedo now considers the most efficient swimsuit yet: the LZR Racer. Surface drag testing was performed with

the help of Langley, and additional water flume testing and computational fluid dynamics were performed with guidance from the University of Otago (New Zealand) and ANSYS Inc., a computer-aided engineering firm.

"Speedo had the materials in mind [for the LZR Racer]," explains Isaac, "but we did not know how they would perform in surface friction drag testing, which is where we enlisted the help of NASA." The manufacturer says the fabric, which Speedo calls LZR Pulse, is not only efficient at reducing drag, but it also repels water and is extremely lightweight. Speedo tested about 100 materials and material coatings before settling on LZR Pulse.

NASA and Speedo performed tests on traditionally sewn seams, ultrasonically welded seams, and the fabric alone, which gave Speedo a baseline for reducing drag caused by seams and helped them identify problem areas. NASA wind tunnel results helped Speedo "create a bonding system that eliminates seams and reduces drag," according to Isaac. The Speedo LZR Racer is the first fully bonded, full-body swimsuit with ultrasonically welded seams. Instead of sewing overlapping pieces of fabric together, Speedo actually fused the edges ultrasonically, reducing drag by 6 percent. "The ultrasonically welded seams have just slightly more drag than the fabric alone," Isaac explains. NASA results also showed that a low-profile

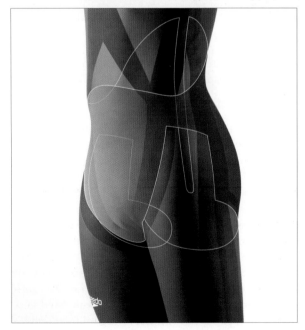

The LZR Racer provides extra compression in key areas to help a swimmer use less energy to swim more quickly.

zipper ultrasonically bonded (not sewn) into the fabric and hidden inside the suit generated 8 percent less drag in wind tunnel tests than a standard zipper. Low-profile seams and zippers were a crucial component in the LZR Racer because the suit consists of multiple connecting fabric pieces—instead of just a few sewn pieces such as found in traditional suits—that provide extra compression for maximum efficiency.

Product Outcome

The LZR Racer reduces skin friction drag 24 percent more than the Fastskin, the previous Speedo racing suit fabric; and according to the manufacturer, the LZR Racer uses a Hydro Form Compression System to grip the body like a corset. Speedo experts say this compression helps the swimmers maintain the best form possible and enables them to swim longer and faster since they are using less energy to maintain form. The compression alone improves efficiency up to 5 percent, according to the manufacturer.

Olympic swimmer Katie Hoff, one of the American athletes wearing the suit in 2008 competitions, said that the tight suit helps a swimmer move more quickly through the water, because it "compresses [the] whole body so that [it's] really streamlined." Athletes from the French, Australian, and British Olympic teams all participated in testing the new Speedo racing suits.

Similar in style to a wetsuit, the LZR Racer can cover all or part of the legs, depending on personal preference and event. A swimmer can choose a full-body suit that covers the entire torso and extends to the ankles, or can opt for a suit with shorter legs above the knees. The more skin the LZR Racer covers, the more potential it has to reduce skin friction drag. The research seems to have paid off; in March 2008, athletes wearing the LZR Racer broke 13 world records. ❖

The LZR Racer reduces skin friction drag by covering more skin than traditional swimsuits. Multiple pieces of the water-resistant and extremely lightweight LZR Pulse fabric connect at ultrasonically welded seams and incorporate extremely low-profile zippers to keep viscous drag to a minimum.

Immersive Photography Renders 360° Views

Originating Technology/NASA Contribution

NASA uses 3-D immersive photography and video for guiding space robots, in the space shuttle and International Space Station programs, cryogenic wind tunnels, and for remote docking of spacecraft. It allows researchers to view situations with the same spatial awareness they would have if they were present. With this type of photography, viewers virtually enter the panoramic image and can interact with the environment by panning, looking in different directions, and zooming in on anything in the 360-degree field of view that is of interest. As the perspective changes, the viewer feels as if he or she is actually looking around the scene, which enhances situational awareness and provides a high level of functionality for viewing, capturing, and analyzing visual data.

Partnership

A **Small Business Innovation Research (SBIR)** contract through Langley Research Center helped Interactive Pictures Corporation (IPC), of Knoxville, Tennessee, create an innovative imaging technology. This technology is a video imaging process that allows real-time control of live video data and can provide users with interactive, panoramic 360° views.

In 1993, the year that the first IPIX camera entered the market, it also received an "R&D 100" award, a prestigious honor given by R&D magazine for significant contributions to the scientific community.

The camera system can see in multiple directions, provide up to four simultaneous views, each with its own tilt, rotation, and magnification, yet it has no moving parts, is noiseless, and can respond faster than the human eye. In addition, it eliminates the distortion caused by a fisheye lens, and provides a clear, flat view of each perspective.

In 1995, an inventor named Ford Oxaal showed the company a technology he had developed which gives users the ability to combine two or more images, whether

Patented IPIX technology produces spherical images that let you feel like you are actually inside the scene.

Image courtesy of Jook Leung.

fisheye or rectilinear, into a single, navigable spherical image. Oxaal convinced IPIX to commercialize this useful showcasing technology, and combined with the advent of the World Wide Web, IPIX was able to execute a successful initial public offering.

The company has changed names at several points along the way. It started out as Telerobotics International, but changed its name to Omniview in 1995 after Oxaal showed his spherical media technology. In 1998, it became Interactive Pictures Corporation, and then later, Internet Pictures Corporation, and finally, IPIX Corporation. In 2007, Minds-Eye-View Inc., founded by Oxaal in 1989 and based in Cohoes, New York, purchased most of the operating assets of IPIX and is now in the process of taking the company and the technology to the entertainment industry. Oxaal is currently president and CEO.

Applications now include what Oxaal calls "homeland reconnaissance," wherein critical infrastructure and public facilities are documented with spherical media; military

reconnaissance; real estate and product showcasing; security and surveillance; and soon, interactive Webcasts.

Product Outcome

Through the NASA SBIR work, IPIX has created two 3-D immersive photography suites: a still image program and a video complement.

The IPIX package is a convenient and powerful documentation and site management tool. It is compatible with many off-the-shelf digital cameras and the final pictures are viewable in any immersive viewing formats, giving users a handful of benefits, including ease of use and the ability to capture and save an entire spherical environment with just two shots. The two images are fused together with no discernable seam, and the viewer can navigate throughout the picture from a fixed location. This is particularly helpful for virtual tours and has been widely embraced by the real estate, hotel, and automobile industries.

With NASA assistance, IPIX developed the first commercially available immersive 360-degree photography, with widespread applicability in real estate and security markets.

IPIX's immersive video suite also offers many benefits. Users can count on immersive video to capture and save digital representations of entire environments, while providing multiple simultaneous views with a single camera and no moving parts. From within the immersive video view, users can electronically pan, tilt, and zoom, while the camera remains motionless. The system also provides wide, complete coverage, with no blind spots, and the files can be transmitted efficiently over networks, even over existing, commercial IP-based platforms.

Both of these camera systems can be employed in virtually any situation where immersive views are needed. They have been used in casinos, airports, rail systems, parking garages, schools, banks, stores, gas stations, automobile dealerships, amusement parks, hotels, homes for sale or rent, cruise ships, warehouses, power plants, incarceration facilities, theaters, stadiums, shopping centers, military facilities, government centers, assisted living centers, hospitals, gated communities, multi-tenant complexes, manufacturing plants, museums, hospitals, office buildings, colleges and universities, courts, and convention centers, to name just a few. Potential applications, however, are limitless.

In 2004, IPIX security cameras were chosen for surveillance of the 2004 Democratic National Convention in Boston and the 2004 Republican National Convention in New York. That same year, the technology was used for surveillance at the 30th G8 Summit at Sea Island, Georgia, and during the President's second inaugural parade in Washington, DC. More recently, the technology has been used to secure everything from the CircusCircus Las Vegas Hotel and Casino to Meade High School at Fort George G. Meade, Maryland, to the Mt. Pleasant, Illinois, City Hall.

The technology isn't only applicable to safety and surveillance uses, though. It is a popular complement to real estate and hotel Web sites, where visitors can take virtual tours of properties online. ❖

Historic Partnership Captures Our Imagination

Famous for its broadcast of greetings on Christmas Eve of 1968 as the crew circled the Moon, Apollo 8 was the first manned mission ever to reach escape velocity. This classic photo of Earth and the lunar surface was taken with a 270-millimeter lens on a Hasselblad camera, and has been heralded as one of the greatest photos of the 20th century by the likes of Time, Life, Sky and Telescope, and other magazines. While frequently printed with the Earth above the lunar surface, the orientation is original and reflects the view of the crew.

Originating Technology/NASA Contribution

Timeless, beautiful, and haunting images: A delicate blue marble floating in the black sea of space; a brilliant white astronaut suit, visor glowing gold, the entire Earth as a backdrop; the Moon looming large and ghostly, pockmarked with sharp craters, a diaphanous grey on deep black. Photographs from space illustrating the planet on which we live, the space surrounding it, and the precarious voyages into it by our fellow humans are among the most tangible products of the Space Program. These images have become touchstones of successive generations, as the voyages into space have illuminated the space in which we live.

In 1962, Walter Schirra blasted off in a Mercury rocket to become the fifth American in space, bringing with him the first Hasselblad camera to leave the Earth's atmosphere, recently purchased from a camera shop near Johnson Space Center in Houston—but not the last. The camera, a Hasselblad 500C, was a standard consumer unit that Schirra had stripped to bare metal and painted black in order to minimize reflections. Once in space, he documented the wonder and awe-inspiring beauty around him, and brought the images back for us to share. The Hasselblad 500C cameras were used on this and the last Project Mercury mission in 1963. They continued to be used throughout the Gemini space flights in 1965 and 1966.

Since then, a number of different camera models have been put to use, but the images taken with the boxy, black Hasselblads have remained true classics. Noted for the amazing sharpness of the photos, the Hasselblads stood up to the rigors of operating in space, facing from -65 °C to over 120 °C in the sun. Many shots have become historic treasures: the first spacewalk during the Gemini IV mission in 1965; the first venture to another celestial body during Apollo VIII, including the iconic "Earthrise" photograph; and the first landing on the surface of the Moon during Apollo XI. These pictures were published

The ELS is a modified 553 ELX, with flash metering removed and leatherette replaced with thin metal plates. This camera was used in the early 1990s on the space shuttle missions. The film magazines use 70-millimeter perforated film and are equipped with electronic data imprinting, enabling the recording of time and picture number for each exposure.

around the world, and have become some of the most recognizable and powerful photographs known.

Several different models of Hasselblad cameras have been taken into space, often modified in one way or another to ease use in cramped conditions and while wearing space suits, such as replacing the reflex mirror with an eye-level finder.

Partnership

Victor Hasselblad AB, of Gothenburg, Sweden, has enjoyed a very long-lived collaboration with NASA. Working primarily with Johnson, the last four decades have seen a frequent exchange of ideas between Hasselblad and NASA via faxes, telephone calls, and meetings both in Sweden and the United States. Initially, most meetings were held at Hasselblad headquarters in Gothenburg, to

Astronaut Buzz Aldrin, lunar module pilot of the first lunar landing mission, poses for a photograph beside the United States flag. Astronaut Neil A. Armstrong took this picture with a 70-millimeter Hasselblad lunar surface camera, one of three Hasselblad cameras the mission carried to the Moon.

Earth resembles a child's marble in this amazing photo, taken during the Apollo 17 mission. NASA officially credits the image to the entire crew, Eugene Cernan, Ronald Evans, and Jack Schmitt, all of whom took photographic images with the onboard Hasselblad.

be as close to the core activities as possible. Since then, collaboration with NASA has allowed what was once a very small company in international terms to achieve worldwide recognition. Hassleblad's operations now include centers in Parsippany, New Jersey; and Redmond, Washington; as well as France and Denmark.

One direct development of this partnership, the 553ELS, is the space version of the 553ELX model, available commercially for years. This camera has adopted several key features and improvements, such as: the fixation of the mirror mechanism was removed from the rear plate to the side walls; aluminum plating replaced the standard black leatherette as the outer covering; the standard 5-pole contact was replaced by a special 7-pole contact equipped with a bayonet locking device; and the battery cover was equipped with a hinge. These changes resulted in increased durability and reliability, and the ELS model has seen frequent use in the shuttle program.

Hasselblad incorporated and refined other modifications by NASA technicians into new models, such as a 70mm magazine developed to meet Space Program needs. Camera modifications included new materials and lubricants to cope with the vacuum conditions outside the spacecraft, and often improved reliability and durability of the cameras. In addition, technicians modified camera electronics to meet NASA's special demands for handling and function, reconstructing lenses and adding large tabs to the focusing and aperture rings to ease handling with the large gloves of an astronaut suit in zero gravity.

Product Outcome

For over four decades, Hasselblad has supplied camera equipment to the NASA Space Program, and Hasselblad cameras still take on average between 1,500 and 2,000 photographs on each space shuttle mission. Just as the remarkable pictures on the surface of the Moon defined an era, the fine pictures of astronauts at work in and around the shuttles and International Space Station (ISS) have helped define the latest era of man's continued exploration of the universe around us.

Likewise, the commercial line of Hasselblad cameras continues to incorporate lessons learned from these voyages. Consumer models have enjoyed such refinements as the revised fixation of the mirror mechanism—the Hasselblad 503CW still features the space-influenced improved mirror mechanism—a design change that gave far better stability for the mirror assembly, and an enlarged exposure button, similar to the one designed for the space models.

In October 2001, the Space Shuttle Discovery, in addition to transporting modules to the ISS, carried a new Hasselblad space camera: a focal-plane shutter camera based on the standard commercial version (203FE) equipped with data imprinting along the edge of the film frame, enabling the recording of time and picture number for each exposure. Since the computers onboard have full control over the position of the shuttle, identification of the exact location captured in a frame has become much easier.

Now that NASA is returning to the Moon and is also looking on to Mars for the next stage of exploration, it is without doubt that Hasselblad cameras will be along to document the voyages for those of us remaining on Earth. The relationship that began in a camera shop in Houston, blossomed on the Moon, and matured on the space shuttle, now prepares to reach new heights. As one more small step for a man and giant leap for mankind approaches, we anxiously await the photographs. ❖

On February 7, 1984, astronaut Robert Gibson used a Hasselblad camera to take this picture of astronaut Bruce McCandless II, when he became the first person to fly untethered in space. McCandless traveled more than 300 feet from the Space Shuttle Challenger, 150 nautical miles above Earth.

Outboard Motor Maximizes Power and Dependability

Originating Technology/NASA Contribution

Developed by Jonathan Lee, a structural materials engineer at Marshall Space Flight Center, and PoShou Chen, a scientist with Huntsville, Alabama-based Morgan Research Corporation, MSFC-398 is a high-strength aluminum alloy able to operate at high temperatures. The invention was conceived through a program with the Federal government and a major automobile manufacturer called the Partnership for Next Generation Vehicles. While the success of MSFC-398 can partly be attributed to its strength and resistance to wear, another key aspect is the manufacturing process: the metal is capable of being produced in high volumes at low cost, making it attractive to commercial markets.

Since its premiere, the high-strength aluminum alloy has received several accolades, including being named Marshall's "Invention of the Year" in 2003, receiving the Society of Automotive Engineering's "Environmental Excellence in Transportation Award" in 2004, the Southeast Regional Federal Laboratory Consortium "Excellence in Technology Transfer Award" in 2005, and the National Federal Laboratory Consortium's "Excellence in Technology Transfer Award" in 2006.

Realizing the potential commercial applicability of MSFC-398, Marshall introduced it for public licensing in 2001. The alloy's subsequent success is particularly apparent in its widespread application in commercial marine products.

Partnership

A worldwide leader in the design, development, and distribution of a wide variety of land and water vehicles, including outboard motors, Bombardier Recreational Products (BRP) Inc., came across a description of the NASA alloy and was immediately intrigued. The Canada-based company decided to meet with NASA in April 2001, to explore how the technology could strengthen its products. BRP and NASA identified an application for

Jonathan Lee, a structural materials engineer at Marshall Space Flight Center was on the team that developed MSFC-398, a high-strength aluminum alloy now being used in high-performance marine outboard engines.

high-performance outboard engine pistons. Prototype production started in July, and the Boats and Outboard Engines Division of BRP, based in Sturtevant, Wisconsin, signed the licensing agreement exactly 1 year later.

"Having a proper mixture of the alloy's composition with the correct heat treatment process are two crucial steps to create this alloy for high-temperature applications," said Lee. "The team at Bombardier worked hard with the casting vendor and NASA inventors to perfect the casting of pistons, learn and repeat the process, and bring its product to market. Chen and I are honored to see something we invented being used in a commercial product in a very rapid pace. We still have to pinch ourselves occasionally to realize that BRP's commercialization effort for this alloy has become a reality. It's happened so quickly."

"The usual cycle for developing this type of technology, from the research stage to the development phase, and finally into a commercial product phase may take several years and more than a $1 million investment," Lee said. In this case, it occurred in fewer than 4 years and at a fraction of that cost.

BRP also applauded NASA for its prompt assistance. "The demands of the outboard engine are more significant than any other engine NASA had ever encountered," claims Denis Morin, the company's vice president of engineering, outboard engines. "The team from NASA was on the fast track, learned all the intricacies, and delivered an outstanding product." BRP incorporated the alloy pistons into a brand new mid-power outboard motor that the company affirms is "years beyond carbureted two-stroke, four-stroke, or even direct injection" engines.

Product Outcome

While a four-stroke engine generally runs cleaner and quieter than its two-stroke counterpart, it lacks the power and dependability; and the two-stroke engine, which generally contains 200 fewer parts than a comparable four-stroke motor, literally has fewer things that can go wrong. Evinrude E-TEC is a line of two-stroke motors that maintain the power and dependability of a two-stroke with the refinement of a four-stroke. The Evinrude E-TEC is also the first outboard motor that will not require oil changes, winterization, spring tune-ups, or scheduled maintenance for 3 years of normal recreational use. It incorporates the NASA alloy into its pistons,

The Evinrude E-TEC line of outboard engines uses a NASA-derived aluminum alloy.

significantly improving durability at high temperatures while also making the engine quieter, cleaner, and more efficient.

The E-TEC features a low-friction design completely free from belts, powerhead gears, cams, and mechanical oil pumps; a "sure-start" ignition system that prevents spark plug fouling and does not require priming or choking; and speed-adjusting failsafe electronics that keep it running even if a boat's battery dies. A central computer controls the outboard engine's single injector, which is completely sealed to prevent air from entering the fuel system and thus minimizes evaporative emissions. Furthermore, the E-TEC auto-lubing oil system eliminates the process of having to mix oil with fuel, while complete combustion

precludes virtually any oil from escaping into the environment. When programmed to operate on specially designed oil, the E-TEC uses approximately 50 percent less oil than a traditional direct injection system and 75 percent less than a traditional two-stroke engine. Additionally, when compared to a four-stroke engine, the E-TEC creates 80 percent less carbon monoxide while idle.

As an added bonus for fishermen, the new piston design also reduces the slapping sound usually made when pistons slide up and down in the engine's cylinder, a sure sign to fish that someone is coming for them with a worm on a hook.

Ranging from 40-horsepower (hp) models to 300-hp models, Evinrude E-TEC engines won the

prestigious "2003 Innovation Award" from the National Marine Manufacturers Association at the annual Miami International Boat Show, and are the only marine engines to have ever received the U.S. Environmental Protection Agency's "Clean Air Technology Excellence Award."

E-TEC also received a testimonial from an individual who put the engine to an incredible test in the most unusual of conditions: While BRP often hears from boaters who depend on its engines in tropical, warm, and temperate climates, the company had heard about an individual from the small Alaskan village of Koyukuk who runs the Yukon River with his Evinrude just about everyday, from break-up of the iced-over body of water to freeze-up during the long Alaskan winter. The nearest "sizable" town is 400 miles upstream from Koyukuk, so the turbid and turgid river serves as the only "highway" on which to acquire goods, tools, and groceries. That's a pretty good vote of confidence. ❖

Evinrude® is a registered trademark, and E-TEC™ is a trademark of Bombardier Recreational Products Inc.

Space Research Fortifies Nutrition Worldwide

Originating Technology/NASA Contribution

I n addition to the mammoth engineering challenge posed by launching a cargo-laden craft into space for a long-distance mission, keeping the crews safe and healthy for these extended periods of time in space poses further challenges, problems for which NASA scientists are constantly seeking new answers. Obstacles include maintaining long-term food supplies, ensuring access to clean air and potable water, and developing efficient means of waste disposal—all with the constraints of being in a spacecraft thousands of miles from Earth, and getting farther every minute. NASA continues to overcome these hurdles, though, and is in the process of designing increasingly efficient life support systems to make life aboard the International Space Station sustainable for laboratory crews, and creating systems for use on future lunar laboratories and the upcoming long trip to Mars.

Ideal life support systems for these closed environments would take up very little space, consume very little power, and require limited crew intervention—these much-needed components would virtually disappear while doing their important jobs. One NASA experiment into creating a low-profile life support system involved living ecosystems in contained environments. Dubbed the Controlled Ecological Life Support Systems (CELSS) these contained systems attempted to address the basic needs of crews, meet stringent payload and power usage restrictions, and minimize space occupancy by developing living, regenerative ecosystems that would take care of themselves and their inhabitants—recreating Earth-like conditions.

Years later, what began as an experiment with different methods of bioregenerative life support for extended-duration, human-crewed space flight, has evolved into one of the most widespread NASA spinoffs of all time.

Partnership

In the 1980s, Baltimore-based Martin Marietta Corporation worked with NASA to test the use of certain strains of microalgae as a food supply, oxygen source, and a catalyst for waste disposal as part of the CELSS experiments. The plan was for the microalgae to become part of the life support system on long-duration flights, taking on a plethora of tasks with minimal space, energy, and maintenance requirements. During this research, the scientists discovered many things about the microalgae, realizing ultimately that its properties were valuable to people not only in space, but here on Earth, as a nutritional supplement. The scientists, fueled by these discoveries, spun off from Martin Marietta, and in 1985, formed Martek Biosciences Corporation, in Columbia, Maryland.

Product Outcome

Now, after two decades of continued research on the same microalgae studied for use in long-duration space flight, Martek has developed into a major player in the nutrition field, with over 500 employees and annual revenue of more than $270 million. The reach of the company's space-developed product, though, is what is most impressive. Martek's main products, life'sDHA and life'sARA, both of which trace directly back to the original NASA CELSS work, can be found in over 90 percent of the infant formulas sold in the United States, and are added to the infant formulas sold in over 65 additional countries. With such widespread use, the company estimates that over 24 million babies worldwide have consumed its nutritional additives.

Outside of the infant formula market, Martek's commercial partners include General Mills Inc., Yoplait USA Inc., Odwalla Inc., Kellogg Company, and Dean Foods Company's WhiteWave Foods division (makers of the Silk, Horizon Organic, and Rachel's brands).

Why would so many people consume these products? The primary ingredient is one of the building blocks of

Shown here is the Skylab food heating and serving tray with food, drink, and utensils. While this represented a great improvement over the food served on earlier space flights, NASA researchers still had plenty of room for progress.

NASA experiments into plant growth for long-duration space flights led to the identification and manufacturing method for a nutritional supplement now found in everyday foods.

health: A fatty acid found in human breast milk, known to improve brain function and visual development, which recent studies have indicated plays a significant role in heart health. It is only introduced to the body through dietary sources, so supplements containing it are in high demand.

The primary discovery Martek made while exploring properties of microalgae for use in long-duration space flights was identifying *Crypthecodinium cohnii*, a strain of algae that produces docosahexaenoc acid (DHA) naturally and in high quantities. Using the same principles, the company also patented a method for developing another fatty acid that plays a key role in infant health, arachidonic acid (ARA). This fatty acid, it extracts from the fungus *Mortierella alpina*.

DHA is an omega-3 fatty acid, naturally found in the body, which plays a key role in infant development and adult health. Most abundant in the brain, eyes, and heart,

it is integral in learning ability, mental development, visual acuity, and in the prevention and management of cardiovascular disease.

Approximately 60 percent of the brain is composed of structural fat (the gray matter), of which nearly half is composed of DHA. As such, it is an essential building block for early brain development, as well as a key structural element in maintaining healthy brain functioning through all stages of life. It is especially important in infancy, though, when the most rapid brain growth occurs—the human brain nearly triples in size during the first year of life. Breast milk, which is generally two-thirds fat, is a chief source for DHA for children, both a testament to the body's need for this substance and an argument for sustainable sources that can be added to infant formula. Studies have shown that adults, too, need DHA for healthy brain functioning, and that the important chemical is delivered through the diet.

DHA is also a key component in the structural fat that makes up the eye, and is vital for visual development and ocular health. The retina, for example, contains a high concentration of DHA, which the body forms from nutritious fats in the diet. With heart tissue, the U.S. Food and Drug Administration has found supporting evidence that DHA consumption may reduce the risk of coronary heart disease.

This important compound, previously only found in human breast milk, and with undeniable nutritional value, is now available throughout the world. It is one example of how NASA research intended to sustain life in space has found its way back to Earth, where it is improving the lives of people everywhere. ❖

life'sDHA™ and life'sARA™ are trademarks of Martek Biosciences Corporation.

Silk®, Horizon Organic®, and Rachel's® are registered trademarks of the WhiteWave Foods Company.

Aerogels Insulate Missions and Consumer Products

Originating Technology/NASA Contribution

Recently, NASA's Stardust mission used a block of aerogel to catch high-speed comet particles and specks of interstellar dust without damaging them, by slowing down the particles from their high velocity with minimal heating or other effects that would cause their physical alteration. This amazing accomplishment, bringing space particles back to Earth, was made possible by the equally amazing properties of aerogel.

Due to its extremely light weight and often translucent appearance, aerogel is often called solid smoke. Barely denser than air, this smoky material weighs virtually nothing. In fact, it holds the world record for being the world's lightest solid—one of 15 records granted it by Guinness World Records. It is truly an amazing substance: able to hold up under temperatures of 3,000 °F. Aerogels have unsurpassed thermal insulation values (providing three times more insulation than the best fiberglass), as well as astounding sound and shock absorption characteristics.

As a class, aerogels, composed of silicone dioxide and 99.8 percent air, have the highest thermal insulation value, the highest specific surface area, the lowest density, the lowest speed of sound, the lowest refractive index, and the lowest dielectric constant of all solid materials. They are also extremely fragile. Similar in chemical structure to glass, though 1,000 times less dense, they are often prone to breaking when handling—seemingly their only drawback—aside from their cost.

Invented nearly 80 years ago, aerogels are typically hard-to-handle and costly to manufacture by traditional means. For these reasons, the commercial industry found it difficult to manufacture products incorporating the material. However, a small business partnered with NASA to develop a flexible aerogel concept and a revolutionary manufacturing method that cut production time and costs, while also solving the handling problems associated with aerogel-based insulation products.

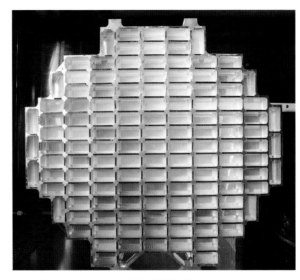

Aerogel has many uses within NASA. For example, it was the perfect lightweight medium for capturing interstellar and comet dust particles during NASA's Stardust mission.

These robust, flexible forms of aerogel can now be manufactured into blankets, thin sheets, beads, and molded parts.

James Fesmire, senior principal investigator at Kennedy Space Center's Cryogenics Test Laboratory, and one of the key inventors of this new technology, says of the advancements, "This aerogel blanket insulation is not only the world's best insulator, but, combined with its favorable environmental and mechanical characteristics, also opens the door to many new design possibilities for buildings, cars, electrical power, and many industrial process systems."

Partnership

Aspen Aerogels Inc., of Northborough, Massachusetts, an independent company spun off from Aspen Systems Inc., rose to the challenge of creating a robust, flexible form of aerogel by working with NASA through a **Small Business Innovation Research (SBIR)** contract with

Kennedy. That contract led to further partnerships for the development of thermal insulation materials, manufacturing processes, and new test methods. This collaboration over many years was a pivotal part for the founding of NASA's Cryogenics Test Laboratory.

Aspen responded to NASA's need for a flexible, durable, easy-to-use aerogel system for cryogenic insulation for space shuttle launch applications. For NASA, the final product of this low thermal conductivity system was useful in applications such as launch vehicles, space shuttle upgrades, and life support equipment. The company has since used the same manufacturing process developed under the SBIR to expand its product offerings into the more commercial realms, making aerogel available for the first time as a material that can be handled and installed just like standard insulation. The development process culminated in an "R&D 100" award for Aspen Aerogels and Kennedy in 2003.

According to Fesmire, "This flexible aerogel insulation idea originated 16 years ago. The problem was to make the world's best insulation material in an easy-to-use form at an affordable price. All these goals have now been achieved through many years of dedicated work."

Product Outcome

Based on its work with NASA, Aspen has developed three different lines of aerogel products: Cryogel, Spaceloft, and Pyrogel. Its work has also infused back into the Space Program, as Kennedy is an important customer.

Cryogel is a flexible insulation, with or without integral vapor barrier, for sub-ambient temperature and cryogenic pipelines, vessels, and equipment. It comes as flexible aerogel blanket insulation engineered to deliver maximum thermal protection with minimal weight and thickness and zero water vapor permeability. Its unique properties—extremely low thermal conductivity, superior flexibility, compression resistance, hydrophobicity, and

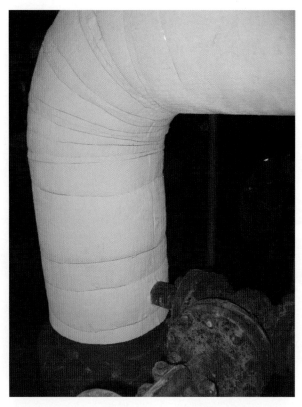

Flexible insulating blankets come in a variety of shapes and sizes, lending themselves to many different applications.

ease of use—make it an ideal thermal protection for cryo-genic applications.

Spaceloft also comes in flexible blanket form and is easy to use. (It can be cut using conventional textile cutting tools, including scissors, electric scissors, and razor knives.) It is designed to meet the demanding requirements of industrial, commercial, and residential applications. Spaceloft is a proven, effective insulator in the oil and gas industries, building and construction, aerospace, automotive, cold chain, and other industries requiring maximum thermal protection within tight space and weight constraints. Spaceloft is used for low-pressure

steam pipes, vessels, and equipment; sub-sea pipelines, hot pipes, vessels, and equipment; footwear and outdoor apparel. Other applications include tents, insulation for interior wall renovation, mobile home exteriors, tractor heat shielding, bus heat shielding, hot water pipes, and solar panels.

Pyrogel is used in medium-to-high pressure steam pipes, vessels, and equipment; aerospace and defense applications; fire barriers; welding blankets; footwear and outdoor apparel. Applications have included insulating an entire polycarbonate plant, a reactor exterior, high-altitude boots, water and gas piping, tubing bundles, yacht exhausts, large vessels, exhaust ducts, ships' boilers, and underground steam lines. The insulation has been proven to be an effective underfoot barrier to extreme cold in the form of insoles for climbers on Mt. Everest, where their light weight and flexibility are also prized. They have even been tested as insoles for ultramarathoners—runners who jog past the 26.2 mile standard marathon distance and sometimes up to 100 miles at a time—who prize the material for its light weight and excellent heat-insulating properties.

It is not just industry and the commercial realm that are benefiting from Aspen's products, though. The work has come full circle, and Aspen is a regular provider of aerogel insulation to NASA, where the material is used on many diverse projects, for space shuttle applications, interplanetary propulsion, and life support equipment. On the space shuttle, it is used as an insulation on the external tank vent's quick-disconnect valve, which releases at liftoff and reaches temperatures of -400 °F. It is also found on the shuttle launch pad's fuel cell systems. At Stennis Space Center's E-3 engine test stand, the blankets are used on the liquid oxygen lines.

In the laboratory, NASA scientists are working to incorporate insulating Aspen aerogels with new polymer materials to create a new category of materials and to create composite foam fillers.

Sponsored by NASA's Space Operations Mission Directorate, engineers are experimenting with the products to create an insulating material that could replace poured and molded foams for a plethora of applications, including in test facilities, on launch pads, and even on spacecraft. ❖

Cryogel™, Spaceloft™, and Pyrogel™ are trademarks of Aspen Aerogels Inc.

Aspen's insulating aerogels have been incorporated into extreme weather gear, where they are prized for both their light weight and extreme insulating performance.

Environmental and Agricultural Resources

NASA's research helps sustain the Earth and its resources.
The technologies featured in this section:

- Locate environmental hazards
- Store clean energy
- Explore the farthest reaches of Earth and space
- Create safer drinking water
- Eliminate emissions
- Put the world at your fingertips

Computer Model Locates Environmental Hazards

Originating Technology/NASA Contribution

Since 1972, Landsat satellites have collected information about Earth from space. Specialized digital photographs of Earth's continents and surrounding coastal regions have helped people study many aspects of our planet and analyze and protect the environment. Resolution and commercial availability of remote sensing through satellite imagery has improved dramatically over the years, and we are now seeing application in a very specific and largely unknown form of environmental threat detection: waste tire piles.

Illegal scrap tire piles are generally considered an environmental hazard because they can become breeding grounds for rodents and mosquitoes and can also generate fumes and toxic fires that are difficult and costly to extinguish. Tires also cause problems in landfills because they settle unevenly and rise to the surface; 38 states ban whole tires from landfills, but most allow tires if they are shredded. Although markets exist for most of the 299 million scrap tires discarded every year in the United States, approximately 40 million scrap tires end up in landfills or in illegal scrap tire piles. In California alone, at least 8 million scrap tires are dumped illegally every year. Although the number of tires in stockpiles has dropped by more than half from the 1994 levels of 700 to 800 million, cleaning up these piles is expensive and, until recently, the illegal piles were especially difficult to locate.

Partnership

Catherine Huybrechts Burton founded San Francisco-based Endpoint Environmental (2E) LLC in 2005 while she was a student intern and project manager at Ames Research Center with NASA's DEVELOP program, which sponsors student internships in applying Earth science data and technology to local policy issues. During that time, the California Integrated Waste Management Board (CIWMB) proposed a pilot project to NASA: develop a new method for mapping waste tire piles.

Becky Quinlan, of Endpoint Environmental LLC, calibrates the TIRe Model by recording the location of known tire piles.

While Burton surveyed environmental agencies for products already available for locating tire piles, her colleague, Becky Quinlan, developed a software model. They determined that neither satellite imagery nor image processing models were being used to locate tire piles.

Using a mapping algorithm called the *Egeria densa* Image Processing Algorithm (EDIPA), which Burton created for her master's thesis, the 2E team created the Tire Identification from Reflectance (TIRe) Model, which algorithmically processes images using turnkey technology to retain only the darkest parts of an image; this allows researchers to find tire piles far more quickly than if they scan images conventionally. "We take that image with the tires and other dark objects and place it on top of the original image," Burton explains. "It is at this point that the highly trained visual image interpreter must determine what is a pile of tires and what is not."

The low reflectivity of tire piles complicates their detection in photographs and makes the process of finding them more difficult due to the fact that they are easily

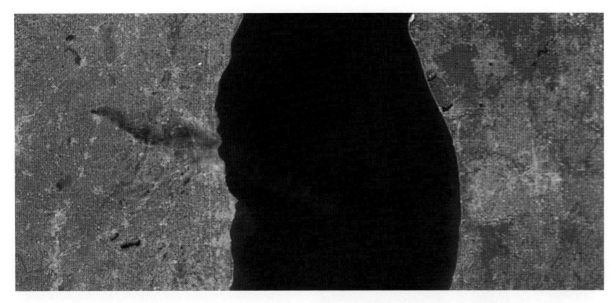

Smoke rises from burning tires on the western shore of Lake Michigan. The TIRe Model processes satellite images to locate illegal tire piles and prevent these environmental hazards.

confused with other dark images in digital photographs. 2E researchers use the TIRe Model to identify specific dark pixels in satellite images, systematically ruling out known features or objects, such as runways, parking lots, shadows, polluted waters, and objects made with recycled tire materials, such as roofs, tubing, and playgrounds. Analysts consider the remaining dark areas suspect, and assume they are probably tire piles, says Burton. The TIRe Model is so finely tuned that it can identify piles containing only 100 tires covering 36 square feet, due to the high resolution of the satellite images.

CIWMB tested the TIRe Model on sites in the coastal area of Sonoma, the desert climate of the Coachella Valley in Southeastern California, and the Lucerne Valley. Without prior knowledge of the locations, 2E located all illegal scrap tire piles targeted by CIWMB, and also located two previously unknown piles. 2E used high-resolution satellite imagery from GeoEye's IKONOS satellite, which

provides 4 meters per pixel, superior to most other satellite imagery that offer 30 meters per pixel.

To improve their analysis, Burton and Quinlan applied specific knowledge of the different climates and of how tire piles were used in specific areas, for instance, as windbreaks, fences, and in gullies to control erosion. 2E did not use thresholding, a common method of separating image features, because the data for tires varies too much between different images. In order to separate tires visually, 2E used a land/water index to eliminate vegetation and water features, and then analyzed hues to eliminate soil and any additional water features. Using these techniques, 2E eliminated 99 percent of the false positives in the satellite images.

After Burton and Quinlan's contracts with NASA concluded, NASA granted 2E's request to share rights to the TIRe Model. 2E has since presented the TIRe Model at several conferences: the American Geophysical Union and the American Society for Photogrammetry and Remote

Sensing (2005); the U.S. Environmental Protection Agency's (EPA) Resource Conservation Challenge at the CIWMB (2006); the Cross-Border Environmental Management Conference at the University of Texas at Austin (2007); the Indiana Geographic Information Council Conference (2007); and the California "CalGIS" Conference (2007).

Product Outcome

Before 2E developed the TIRe Model, methods available for identifying tire piles included aerial and spectral video analysis, but no formal commercial models were available. The TIRe Model now helps state governments and private industry to locate and monitor tire piles, whether legal or illegal. Many states have already cleaned up large numbers of stockpiles, which is partly due to the growing market for reprocessed tire products (such as tire-derived fuel, civil engineering, and rubber-modified asphalt). By the end of 2005, according to the most recent report from the U.S. Environmental Protection Agency (EPA), seven states contained 84 percent of all scrap tires: Colorado, New York, Texas, Connecticut, Alabama, Michigan, and Pennsylvania.

In 2007, 2E worked on a project with San Francisco State University to map tire sites in Northern California and along the California-Mexico border. Because many tires are disposed of illegally by Americans in Mexico, there is a fair amount of interest in locating tire piles along the border. In addition to offering their tire mapping services to California, Indiana, and the EPA, Endpoint Environmental is also negotiating with a private company in Wyoming. Several industry publications have published articles on 2E's tire finding services. 2E is also using Burton's original EDIPA model and the TIRe Model to locate invasive plant species. ❖

Battery Technology Stores Clean Energy

Originating Technology/NASA Contribution

Affordable and reliable clean energy has been a tantalizing, but elusive, quarry. Featured in *Spinoff* 1985 and pioneered by Lawrence Thaller at Lewis (now Glenn) Research Center in the 1970s as a potential alternate energy source for long-term space flight, iron-chromium redox energy storage systems are a hybrid technology that offers the extended support of fuel cells with the flexibility of batteries. They act as "electron buckets" for existing clean energy sources, such as solar or wind, to store and deliver power predictably when needed.

In iron-chromium redox systems, electricity is generated when pumps move the electrolytes into separate sections of a reaction chamber. Electrodes collect that charge, and the electrolytes can then be recharged from an outside power source. Thaller's initial design was a 1-kilowatt system (2 kilowatts at its peak), which used acidified chromium and iron in its solution and relied on soluble redox couples and an ion exchange membrane to generate and store energy in a liquid electrolyte solution.

Partnership

Headquartered in Fremont, California (with offices in Gurgaon, India), Deeya Energy Inc. is now bringing its iron-chromium hybrid flow batteries to commercial customers around the world. Thaller supported Deeya's founder, Saroj Sahu, providing development assistance for the company's proprietary liquid-cells (l-cells.) The l-cells have higher power capability (3 kilowatts) than Thaller's original design, and in January 2008, the Space Foundation approved the l-cells as a Certified Space Technology, a designation for products made possible by space research and development.

Product Outcome

According to Rick Winter, an engineer and vice president with the company, Deeya's l-cells offer a few fundamental differences from the original redox system. "With the advent of modern plastics, we have been able to replace critical components, dramatically improving the system's performance, cost, and life," Winter explains. "We have improved the reliability and reduced the component count and cost so that it can be commercially competitive." Deeya l-cells are effectively 3 times less expensive than lead-acid batteries and 10 to 20 times less expensive than nickel-metal hydride batteries, lithium-ion batteries, and fuel cell options. The system represents a clean energy technology with no poisonous or expensive metals or fumes release.

Like the original redox system, l-cells offer several advantages over traditional lead-acid batteries: lower cost, longer life, small space needs, and excellent performance at high ambient temperatures. Because l-cells, according to Winter, "actually enjoy sitting in the sun," rural communities in India with power supply problems have expressed interest in the technology. Deeya has tested the l-cells in air temperatures up to 120 °F, and offer significantly better performance than lead-acid, which only perform reliably in moderately cool temperatures. L-cells also have the ability to operate for thousands of discharge cycles without boost charging, as opposed to the current generation of rechargeable batteries, which are only good for 200 to 500 deep discharge cycles.

Deeya l-cells store energy within the electrolyte itself, with no solid materials, such as lead-oxide, required. This approach completely decouples the power and energy ratings of the systems. As such, system power is defined

Load-leveling liquid redox systems originated as potential energy storage for long-term space flight.

The size of large refrigerators, Deeya Energy's l-cells can provide backup power to standard cell towers for 4 hours.

"With today's intense interest in energy independence and renewable energy sources," Winter says, "we should expect to see many full-scale commercial products changing the energy landscape in the next few years."

According to Winter, Deeya flow batteries can support a standard cell tower for 4 hours with a unit that is "the size of a large refrigerator." Deeya's customers include cell phone providers in India who need smaller backup systems for their cell towers. Because of the backup l-cell battery systems, customers see improved reliability of services, including fewer dropped calls and fewer power outages. Other uses for l-cells could include backup systems for cash machines and traffic lights.

Another large-scale application can be found on King Island, near Australia's Tasmania, where residents installed similar technology to supplement their wind turbine energy farm; they saw a significant reduction in power outages and fuel costs. Flow batteries reduced the island's carbon dioxide emissions by 2,000 tons per year.

Deeya plans to install many more systems in rural areas in the developing world to provide for improved communications and significant emissions reductions. Plans include "power-station-in-a-box" products for village electrification, combining solar and wind generation sources. Multimegawatt systems will then be developed for large-scale grid-connected applications, since flow cells can improve the operational efficiency and emissions of coal- and gas-fired power plants.

These customers also appreciate the fact that it is easy to increase capacity—by adding more electrolytes—at a relatively low cost. In the long term, Deeya plans to expand into large 50-megawatt batteries. "The bigger the tank, the longer you can support a load," Winter explains. Pacific Gas and Electric Company and Southern California Edison are both exploring use of these large electricity storage applications using flow batteries for backup systems to wind turbines. ❖

by the size of the electrode stacks, and available energy depends only on the size of the electrolyte tanks. Due to the low cost of the active materials, the technology lends itself to applications needing extended support times. Deeya l-cells also have no harmful metals or fumes and are completely recyclable, making them a clean replacement for lead-acid batteries and diesel generators. The most impressive difference between Deeya l-cells and lead-acid batteries, however, is the life expectancy. Under standard use, lead-acid batteries tend to need replacement every 18 months, whereas Deeya l-cells need refurbishing only every 7 years after which they can last indefinitely. In heavy use applications, they are expected to provide three times the life of lead-acid batteries.

Deeya is now building the "smallest flow batteries in the industry," focusing on 2-kilowatt applications. Flow batteries, a form of rechargeable battery in which electrolytes flow through a power cell, are often used in load leveling for clean technologies such as wind and solar power that sometimes have intermittent drops in power. Such systems need backup storage methods to even out the high and low periods of demand and energy generation.

Robots Explore the Farthest Reaches of Earth and Space

Originating Technology/NASA Contribution

"We were the first that ever burst/Into that silent sea," the title character recounts in Samuel Taylor Coleridge's opus *Rime of the Ancient Mariner*. This famous couplet is equally applicable to undersea exploration today as surface voyages then, and has recently been applied to space travel in the title of a chronicle of the early years of human space flight ("Into That Silent Sea: Trailblazers of the Space Era, 1961-1965"), companion to the "In the Shadow of the Moon" book and movie. The parallel is certainly fitting, considering both fields explore unknown, harsh, and tantalizingly inhospitable environments.

For starters, exploring the Briny Deep and the Final Frontier requires special vehicles, and the most economical and safest means for each employ remotely operated vehicles (ROVs). ROVs have proven the tool of choice for exploring remote locations, allowing scientists to explore the deepest part of the sea and the furthest reaches of the solar system with the least weight penalty, the most flexibility and specialization of design, and without the need to provide for sustaining human life, or the risk of jeopardizing that life.

Most NASA probes, including the historic Voyager I and II spacecraft and especially the Mars rovers, Spirit and Opportunity, feature remote operation, but new missions and new planetary environments will demand new capabilities from the robotic explorers of the future. NASA has an acute interest in the development of specialized ROVs, as new lessons learned on Earth can be applied to new environments and increasingly complex missions in the future of space exploration.

Partnership

Deep Ocean Engineering (DOE) Inc., of San Leandro, California, designs and manufactures a complete range of underwater ROV systems. Founded in 1982, DOE has over 25 years of continuous operating experience in providing a wide range of innovative and cost-effective robotic solutions. Over 500 ROV systems have been designed, built, and delivered to hundreds of customers in over 30 countries.

During the 1990s, DOE received the first of several **Small Business Innovation Research (SBIR)** contracts from NASA to develop ROV technologies. Since many of those technologies are similar to what astronauts might use in space, the firm first worked on an ROV that could test enhanced human interfaces through an SBIR contract with Ames Research Center.

NASA recently followed this work with a Phase II SBIR contract, exploring the research and development of a versatile apparatus and method for remote robot mobility. Robotics intended for exploration of extremely remote and harsh environments must be extraordinarily versatile as well as robust. They must be able to perform in and adapt to unknown and highly variable physical surroundings—robots with tracks, wheels, or articulating legs can provide effective mobility on certain kinds of terrain. However, these same robots would likely fail if their mission requirements included roving over roughly contoured surfaces or in water, sand, slush, or ice.

DOE engineers developed a concept for a versatile and robust locomotion methodology based on snake and worm morphologies. This "super snake" has the ability to transition seamlessly from one environment to another, such as land to water, to burrowing into soft sediment.

Field experiments with this robot include sites such as San Francisco Bay, California's Mono Lake, and desert environments such as Death Valley. The prototype equipment will be used by NASA on future projects, such as scientific research in the Dry Valley lakes of Antarctica,

Deep Ocean Engineering Inc. provides robotic solutions for various underwater applications in harsh and diverse operating environments. DOE has been in continuous operation for 25 years and has sold more than 500 ROV systems in over 30 countries worldwide.

and experiments to evaluate concepts for exploration of Mars and Jupiter's moon, Europa.

Product Outcome

The Monterey Bay Aquarium Research Institute (MBARI) has teamed with the National Oceanic and Atmospheric Administration on an Antarctic research project to study chemical and biological characteristics of icebergs. Because of the danger involved in getting too close to the icebergs, the team has been using a Deep Ocean DS2 ROV. The DS2 (two motor) ROV started life as a delivery to NASA in 1998, and has been on loan to MBARI. The ROV is in the process of being modified to become a DS4 (four motor), including additional thruster/motors and additional flotation.

Michael Gilson, vice president of Customer Solutions for DOE, explained the members of the Phantom family of ROVs: "The DS2 is a deeper-diving version of the S2 and the DS4 is a deeper version of the S4," he said. "Unlike [with] a diver, depth is not a limiting factor, nor is water temperature or clarity of the water."

Configured submersible delivery systems also integrate a wide variety of sensors, tools, electronic navigational controls and tracking systems, instrumentation packages, and accessories.

The DS2 has an array of sensors that includes two suction samplers, a water sampler, plankton net, dissolved oxygen sensor, water speed sensor, two cameras (SIT and HD), three-function manipulator, and a tracking system. The Phantom family includes more than 70 deliveries in many countries for applications including military, archeology, science, oil and gas, and ship hull inspection.

Once the ROV is deployed in the water, the operator controls it and the attached instruments from the surface. From helping to recover drowning victims to inspecting off-shore oil platforms and gathering intelligence, DOE's ROVs have many underwater uses. The U.S. armed forces use the ROVs in security measures and intelligence gathering, and Hydro Quebec—Canada's largest electric utility and second largest corporation, which generates more than 95 percent of its production from hydroelectric facilities—uses a DOE ROV to inspect dams and hydroelectric apparatus for damage, saving an estimated $10 million in two seasons. More than 40 universities and scientific organizations conduct marine studies with the ROVs.

After more than 3 years of engineering design and development, including extensive discussions with customers worldwide, DOE launched the VectorROV product family. The state-of-the-art Vector family has three models to meet unique customer requirements. For example, the Vector L4 has been designed to meet the requirements of military, maritime, power generation, offshore petroleum, and scientific applications, and combines superior power, telemetry, and payload with ease of use, ruggedness, and reliability.

The power and control system for the Vector L4 is designed for simplicity and ease of use, with multiple microprocessors providing redundancy and expanded capabilities. The design incorporates intuitive, computer-aided, always active diagnostics facilitating maintenance of the system in the harshest environments by technicians with a minimum of training. Specifically designed high-performance brushless thrusters provide the highest

DOE's ROV systems have been utilized in a broad range of industry applications, including military, security, salvage, long tunnel and pipeline inspection, customs, and nuclear and hydroelectric power plants.

power-to-weight ratio and reliability compared to other vehicles in its class. The Vector L4 graphical user interface with multiple menu screens provides intuitive feedback and active user control for ease of vehicle handling, navigation, collection and display of sensor data, as well as setting and storing custom system configurations.

Fabricated using modern marine-grade aluminum and composite materials, the chassis is modular with quick access to all the parts for ease of servicing and replacement as required. Constructed from polypropylene, the chassis is resilient, non-corroding, and maintenance free. Ancillary equipment is easy to mount and integrate. ❖

Vector™ is a trademark of Deep Ocean Engineering Inc.

Portable Nanomesh Creates Safer Drinking Water

Originating Technology/NASA Contribution

Providing astronauts with clean water is essential to space exploration to ensure the health and well-being of crewmembers away from Earth. For the sake of efficient and safe long-term space travel, NASA constantly seeks to improve the process of filtering and re-using wastewater in closed-loop systems. Because it would be impractical for astronauts to bring months (or years) worth of water with them, reducing the weight and space taken by water storage through recycling and filtering as much water as possible is crucial. Closed-loop systems using nanotechnology allow wastewater to be cleaned and reused while keeping to a minimum the amount of drinking water carried on missions.

Current high-speed filtration methods usually require electricity, and methods without electricity usually prove impractical or slow. Known for their superior strength and electrical conductivity, carbon nanotubes measure only a few nanometers in diameter; a nanometer is one billionth of a meter, or roughly one hundred-thousandth the width of a human hair. Nanotubes have improved water filtration by eliminating the need for chemical treatments, significant pressure, and heavy water tanks, which makes the new technology especially appealing for applications where small, efficient, lightweight materials are required, whether on Earth or in space.

"NASA will need small volume, effective water purification systems for future long-duration space flight," said Johnson Space Center's Karen Pickering. NASA advances in water filtration with nanotechnology are now also protecting human health in the most remote areas of Earth.

Partnership

In 2003, Seldon Laboratories LLC, of Windsor, Vermont, received their first NASA **Small Business Innovation Research (SBIR)** award for a "Nanomechanical Water Purification Device." In Phase I, Seldon designed a filtration system using its proprietary filters with low-energy and low-space requirements, using carbon nanotubes to reduce the power requirements of closed-loop water (and other liquid) treatment systems and to eliminate hazardous chemical treatments.

Seldon patented a media for a lightweight, low-pressure water purifier that used carbon nanotubes to remove microorganisms from large quantities of water

Research into providing clean drinking water for astronauts like Stephanie Wilson, STS-121 mission specialist, has yielded new water filtration products on Earth.

quickly; the production process fused the nanotubes into a membrane. After 2 years of testing, in-house data indicated that the Seldon filters reliably removed waterborne viruses and bacteria. Each cylindrical filter cleaned water successfully for 40-60 days, surpassing the NASA requirements for 30-day missions.

Phase II of the SBIR agreement included treating higher volumes of water with larger membranes; carbon nanotubes were connected into a mesh. Engineers studied filter regeneration and bio-film abatement techniques for extending filter life, and subjected the filter elements to chemical removal testing. Although Seldon designed the membranes for bio-removal (including viruses and bacteria), the filters also removed some contaminants and chemicals—Seldon scientists affirm specific chemical adsorption can be engineered into the nanotubes for future projects.

Testing in U.S. Environmental Protection Agency-certified facilities showed that Seldon's Nanomesh removed microorganisms such as bacteria (99.9999 percent, or 6 log), viruses (99.99 percent, or 4 log), *Cryptosporidium parvum*, *Giardia lamblia*, and chemical contaminants including arsenic, lead, benzene, copper, dioxins, herbicides, mercury, and endotoxins, such as *Escherichia coli* (*E.coli*) and *Salmonella*.

Seldon designed, constructed, and tested prototypes of the final Nanomesh filter, before delivering them to NASA in November 2006. Seldon's filters have potential applications in industrial water purification systems, industrial decontamination, and commercial and household use.

Product Outcome

The commercial version of the carbon Nanomesh designed under the NASA SBIR agreement was released as the WaterStick. Operating as simply as a straw and almost as small, the WaterStick cleans about 5 gallons (200 milliliters) of water per minute simply using water pressure (up to 3 PSI) and gravity: without electricity,

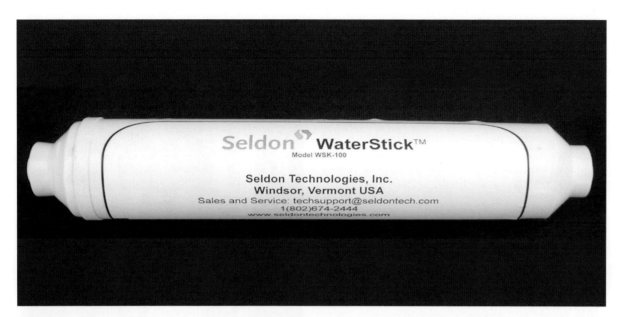

Comparable in size to a zucchini and operated like a straw, Seldon's WaterStick cleans water by using nanotechnology to remove sediment, chemicals, bacteria, viruses, and microorganisms.

heat, environmental impact, or chemical additives (such as iodine or chlorine used by other filters).

The lightweight filters measure about 10½ inches by 2 inches (27 cm by 5 cm), or similar in size to a zucchini. The WaterStick also removes chlorine and iodine, thereby creating water that is not only potable, but palatable as well. Lastly, the WaterStick is easy to use, requiring only a simple filter change every 7 weeks (for one person) or 73 gallons (275 liters.) The manufacturer considers the WaterStick to be "perfect for short-duration uses where other solutions are impractical."

The ease and portability of the WaterStick makes it useful for a variety of applications in which access to clean water and electricity are restricted, such as remote locations or disaster areas. The Seldon WaterStick weighs only 8 ounces and is sold as a standalone device through which water can be poured. Seldon is currently targeting the lightweight WaterStick to disaster relief workers,

researchers, and military personnel in remote locations with no wastewater facilities; future markets may include recreational hikers.

In locations where waterborne illness can prove disastrous, clean water is even more essential. The water filtration element achieves a high level of bacteria and virus removal while maintaining high flow rates and low pressure drops across the filter's surface. According to Seldon CEO Alan Cummings, "Seldon products containing Nanomesh can provide mobile, life-saving ground and surface water filtration in field and disaster relief environments by removal of microorganisms that cause waterborne diseases."

Most of the other commercial water filters do not offer the same level of filtration as the WaterStick; those employing a charcoal or ceramic filter are often less effective against viruses, and those offering comparable protection employ chemicals. Alternately, products that

offer a high level of filtration may use ultraviolet light and thereby require batteries or electricity. Ultraviolet water treatment works well, but tends to be less effective in highly contaminated water with a high percentage of particulates. Seldon engineer Jonathan Winter states, "Our technology is superior, because it doesn't need a certain amount of time or electricity."

Seldon plans to develop an under-the-sink water filtration unit for residential use; like the WaterStick, these larger units would use water pressure to push water through the Nanomesh, and would not require electricity. Other target areas for development may include industrial purification, decontamination, and desalination. ❖

Seldon WaterStick™ and Nanomesh™ are trademarks of Seldon Technologies Inc.

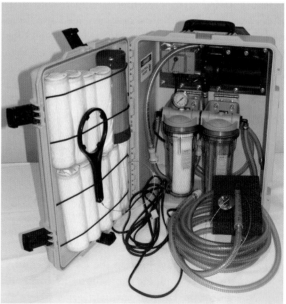

Recently used in Rwanda, Seldon's WaterBox is the size of a small suitcase and is capable of supporting the water needs of a small village.

Innovative Stemless Valve Eliminates Emissions

Originating Technology/NASA Contribution

The past, present, and future of NASA launch and space travel technologies are steeped in the icy realm of cryogenics. NASA employs cryogenics, the science of generating extremely low temperatures and the behavior of materials at those temperatures, in a variety of fluid management and low-temperature applications including vehicle propulsion, high-pressure gas supply, life support equipment, food preservation and packaging, pharmaceutical manufacturing, and imaging devices. Most prominently, cryogenic liquid hydrogen fuel is used by the space shuttle as its primary means of achieving orbit.

Looking to the future, NASA's Constellation Program, which is focused on developing next-generation launch vehicles for planned trips to the Moon, Mars, and beyond, is incorporating cryogenics into the upper stage of the Ares I crew launch vehicle, the core stage and Earth departure stage of the Ares V cargo launch vehicle, and other systems. In support of this work, NASA's Cryogenic Fluid Management (CFM) Project is performing experimental and analytical evaluation of several types of propellant management systems to enable safe and cost-effective exploration missions. The CFM Project is led by Glenn Research Center, with support from Marshall Space Flight Center, Johnson Space Center, Ames Research Center, Kennedy Space Center, and Goddard Space Flight Center.

Cryogenic propellants have been favored for their high-energy, high-efficiency performance; however, cryogenic propellants have not been used in extended-duration space missions since they are difficult to maintain in their highly dense liquid form at the low temperatures common in space and on the Moon. Performance requirements for the Earth departure stage, as well as the lunar lander descent and ascent stages, point toward the use of cryogenic engines and propellants for missions of up to 210 days on the surface of the Moon.

CFM team research is focused on the storage, fluid distribution, liquid acquisition, and mass gauging of cold propellants. These tasks will reduce the development risk and increase the ability of advanced subsystems to store and distribute cryogenic propellants required for long-term exploration missions. CFM utilizes the development of prototype CFM hardware, the creation and use of analytical models to predict subsystem performance, and the execution of ground-based tests using liquid oxygen, liquid hydrogen, and methane to demonstrate the performance, applicability, and reliability of CFM subsystems.

Partnership

Big Horn Valve Inc. (BHVI), of Sheridan, Wyoming, won a series of **Small Business Innovation Research (SBIR)** and **Small Business Technology Transfer (STTR)** contracts to explore and develop a revolutionary valve technology based on cryogenically proven Venturi Off-Set Technology (VOST). In 2001, BHVI first worked with Kennedy on an SBIR contract, "New and Innovative Valving Technology for Cryogenic Applications." In 2003, BHVI's proposal, "Low-Mass VOST Valve," was selected from a field of 2,696 other entries nationwide to receive a Phase I SBIR contract, sponsored by Marshall under the Next Generation Launch Technology Program. This project developed a low-mass, high-efficiency, leak-proof cryogenic valve using composites and exotic metals, and had no stem-actuator, few moving parts, with an overall cylindrical shape. The valve geometry reduces launch vehicle complexity and facilitates assembly and testing. This valve also enhances reliability and safety, due to the inherent simplicity and leak-proof characteristic of the design.

According to principal investigator Zachary Gray, the work with NASA helped BHVI gain " . . . a lot of experience with extreme environments. We gained a lot of contacts in the aerospace and cryogenic community. By attempting to solve NASA's unique problems, we have greatly simplified the valve design while at the same time

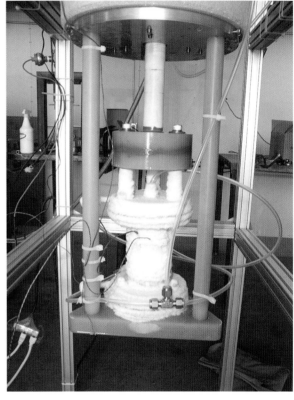

The quick-disconnect valve is shown here undergoing cryogenic testing as part of the Phase II SBIR contract with Johnson Space Center.

demonstrating that VOST worked well from half-inch diameters up to a 4-inch diameter."

In 2005 and 2006, BHVI continued this work upon receipt of a Phase II SBIR award for "Low-Mass VOST Valve" from Marshall, and two Phase II SBIR awards from Johnson. These projects, "In-Space Cryogenic VOST Connect/Disconnect," and "VOST Flow-Control Valve," demonstrated that VOST cryogenic flow control offered precise linear flow control across the entire dynamic range, held its position without power, and required low actuation energy. This project paved the way

A 4-inch valve undergoing cryogenic testing in a tank as part of the SBIR work with Marshall Space Flight Center.

for VOST valve application in future spaceport systems, advanced cryocoolers, launch vehicles, and high-pressure flow-control valves.

Product Outcome

The precise control, inherent simplicity and durability, and demonstrated abilities of the VOST system afford many commercial applications, including petroleum refining, specialty chemical and high-purity pharmaceutical production, and the manufacture of industrial flow-control valves and food processing equipment. The VOST design is a magnetically actuated system in which internal magnets are used to close, open, and throttle the valve. This innovative stemless design, the only one commercially available, is emission-free with no external leakage of vapors or fluid and allows for superior fluid handling features (such as throttling, low-pressure drop, and axial envelope) within a single device structure. The VOST system has potential application in all valving environments and represents a new concept for a tradition-bound industry.

BHVI enlisted the help of Moog Inc. to develop a pneumatically operated control valve. This photo shows a 4-inch valve undergoing cryogenic testing.

In December 2003, BHVI was selected to exhibit the VOST technology at the 2004 World's Best Technologies Showcase, in Dallas, Texas. The 75 exhibitors selected came from the Nation's most advanced research facilities, top universities, and entrepreneurs. The technologies displayed are considered the best of the best.

The first commercial MagVOST was installed March 16, 2006, at Windsor Energy Inc.'s methane coal gas field, east of Kaycee, Wyoming. Future applications are expected to include in-flight refueling of military aircraft and high-volume gas delivery systems such as liquefied natural gas (LNG). Big Horn is also exploring opportunities that require extreme attention to safety, such as with hydrofluoric acid in the petroleum refining industry and in the nuclear industry, and received an SBIR contract from the U.S. Navy to develop a bi-directional VOST valve for use as an isolation valve on ships. ❖

VOST™ and MagVOST™ are trademarks of Big Horn Valve Inc.

Web-Based Mapping Puts the World at Your Fingertips

Originating Technology/NASA Contribution

NASA's award-winning Earth Resources Laboratory Applications Software (ELAS) package was developed at Stennis Space Center. Since 1978, ELAS has been used worldwide for processing satellite and airborne sensor imagery data of the Earth's surface into readable and usable information. In addition to satellite applications such as data from Earth-observing SPOT (Satellites Pour l'Observation de la Terre) satellites, ELAS was applied to aircraft data and medical imagery. While the ensuing decades have seen great improvements in software and imaging technologies, the original developers of ELAS had the foresight to use a modular design, allowing capabilities to be added and expanded as the remote-sensing industry grew.

"ELAS provided a dictionary of parameters used consistently in over 100 applications, which aided users greatly. In addition, ELAS modules typically used a common set of basic commands; after a short introduction, the only difficulties with using ELAS were discovering what the various modules could do," reflected Dr. Ray Seyfarth, one of the original developers of ELAS.

ELAS could be considered an "All-Star" NASA-derived technology, having made frequent appearances in *Spinoff* in myriad applications, including use by NASA's Technology Application Center in studies of the urban growth in the Nile River Delta (*Spinoff* 1985); Delta Data Systems in the construction of their proprietary ATLAS geographic information system (*Spinoff* 1986) and Advanced Geographic Information System (*Spinoff* 1993); Ducks Unlimited Inc., in the construction of its Habitat Inventory and Evaluation Program (*Spinoff* 1987); Medical Image Management Systems in its diagnostic aid and image storage and distribution MD Image System (*Spinoff* 1991); Martin Marietta and the Mid-Atlantic Remote Sensing Center in the development of the Integrated Automated Emergency Management Information System earthquake preparedness program

This image was constructed for a wetland mitigation study utilizing aerial imagery and soils and elevation data.

(*Spinoff* 1991); and DATASTAR Inc., in the DATASTAR Image Processing Exploitation (DIPEx) desktop and Internet image processing, analysis, and manipulation software (*Spinoff* 2003), the development of which has now been continued in DIPEx Version III.

Partnership

In 1992, Stennis' Commercial Technology Program made ELAS available under the Freedom of Information Act, which allows federally developed technologies that are not patent protected to be transferred to U.S. companies. In DIPEx Version III, DATASTAR Inc., of Picayune, Mississippi, has once again used ELAS software to bring a

tool to the public that captures and expands the abilities of ELAS. Improvements in the quality of satellite data have demanded corresponding development of processors, and DATASTAR leveraged the original ELAS design to address today's local and regional database requirements.

Product Outcome

DIPEx is now a mature, user-friendly application used to perform image processing, analysis, and to manipulate remotely sensed imagery data. DIPEx can separate and provide data classifications, false color composites, soils, corridor analysis, subsurface vegetation, data enrichment, mosaics, and geographical information systems (GIS). The architecture of DIPEx allows a wide range of scalability, and the dynamic dimensionality of DIPEx internals assures that the software is current with leading-edge computer hardware.

Users request either a deliverable product from DATASTAR or access data sets via their own computers. Web customers subscribe to a selection of data points, then log on and manipulate the data on a secure server which DATASTAR provides to protect the personal data of subscribers. The system is structured to allow hundreds of users to access and extract layers of information simultaneously.

These layers are composed of the combination of a data source and a rendering asset, and are stored under a map view; any number of folders can be used to organize layer information. Map view assets allow a user to specify and save information, including legend, scalebar, geography, output format, and the layers to be rendered. Map views also provide an easy mechanism to share maps over the Web among groups of users. By decoupling the data source and the renderer, the storage and management of the data source are completely separate from the rendering, so a user can use one data source in many layers with different renderings.

The Web interface itself is a significant upgrade for DIPEx Version III. The original interface was composed of

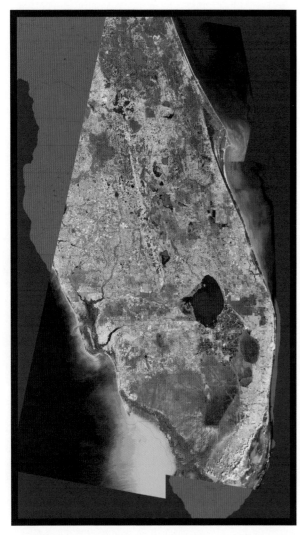

The Florida peninsula citrus growing region, as imaged by Landsat.

HTML pages on which the user posted form information. This architecture was effective, but did not lend itself to reusability and was quickly approaching its limits. Version III is completely implemented using SOAP Web Services.

The Web Services have also proved very useful for other applications, and DATASTAR currently has Microsoft .Net connection software and Perl applications exploiting functionality of the DIPEx server.

A true World Wide Web application that runs using hypertext transfer protocol (HTTP) and starts without a Web browser, Version III evolved with worldwide geospatial dimensionality and numerous other improvements. Version III is difficult to distinguish from a Windows-based application, with all the familiar menu systems, mouse interaction, and drag and drop interfaces. DATASTAR is enhancing the system's mapping capabilities and colorizing data to give it depth. Data provided by DIPEx is compatible with all GIS software packages, including ArcView, ENVI, and ERDAS IMAGINE.

The flexibility and adaptability of the DIPEx system continues a defining trait that has held since the original ELAS was developed. Taking complicated sets of data and integrating them into a clear and useful product has long been the purpose of this software, and Seyfarth enjoys how far the software he helped create has come, affirming, "I am happy to hear that ELAS is alive in the software of DATASTAR. The work from 30 years ago is still valuable; I suspect there is a hold-out somewhere who is still typing ELAS commands."

Dr. Ramona Pelletier Travis, who worked with ELAS as a research scientist in the 1980s and is now the manager of the Innovative Partnerships Program at Stennis, concludes: "ELAS was a fantastic tool then, and I'm glad that its various progeny have seen so much success, including DIPEx. It has been a great example of good government research spinning off to benefit the private sector in a significant way over a long period of time." ❖

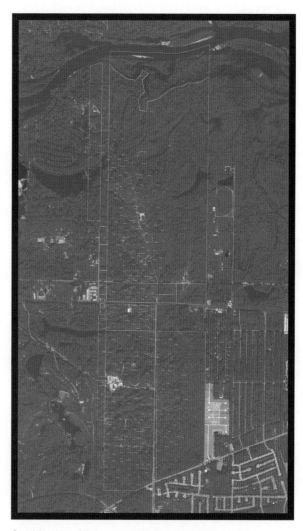

Combination of imagery, point, line, and polygon data.

NASA's work in advanced computing has led to many innovations. The technologies featured in this section:

- Assist satellite designers
- Simplify circuit board manufacturing
- Schedule missions, aid project management
- Analyze complex systems in real time
- Handle image data
- Offer a wide perspective
- Simulate sight
- Monitor tasks
- Bring Main Street to life
- Overcome harsh environments
- Improve network efficiency

Program Assists Satellite Designers

Originating Technology/NASA Contribution

Managed by Goddard Space Flight Center, the Rossi X-ray Timing Explorer (RXTE) was launched on December 30, 1995, from Kennedy Space Center, and to this day, it is still active. The satellite carries several instruments and is part of the Science Mission Directorate's study of deep space. The RXTE measures the timescale of flickering X-rays, called oscillations, revealing the underlying physics of the violent environment around objects such as neutron stars and black holes. The oscillations reveal the nature of the physical environment of the star system, so by studying these oscillations and tracking the same X-ray sources for years, RXTE scientists form a picture of the events that are taking place.

One of the enabling technologies created for the RXTE mission was the Advanced System for Integration and Spacecraft Test (ASIST) software, a real-time command and control system for spacecraft development, integration, and operations. It was designed to be fully functional across a broad spectrum of satellites and instrumentation, while also being user friendly.

Partnership

Annapolis, Maryland-based designAmerica Inc. (DAI), a small aerospace company specializing in the development and delivery of ground control systems for satellites and instrumentation, was one of the organizations that assisted Goddard in the development of the ASIST software.

Realizing that the technology had broader applications in the commercial sector, designAmerica sought assistance from the Innovative Partnerships Program at Goddard, which assisted the company in licensing the software for commercial applications. The company is now marketing ASIST as a commercial-off-the-shelf solution in an arena once restricted to costly, custom-developed, project-specific software.

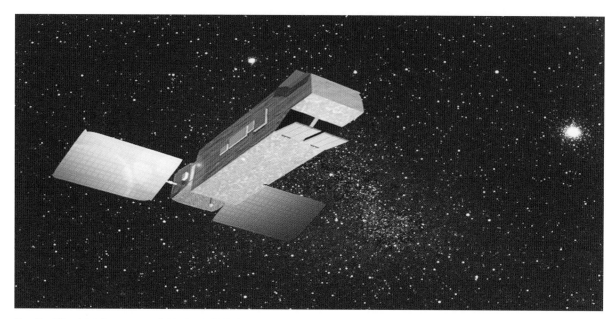

The Far Ultraviolet Spectroscopic Explorer (FUSE) is a NASA-supported astrophysics mission that was launched on June 24, 1999, to explore the universe using the technique of high-resolution spectroscopy in the far-ultraviolet spectral region. Maryland-based designAmerica Inc. provided software to aid this mission.

Product Outcome

ASIST is now a highly complex computer application designed to both meet the demanding technical requirements of modern satellites and their instrumentation and to serve as a tool for the engineers who construct and sustain the technology. The commercial version boasts the same level of functionality as the NASA-developed one, plus adds an approachable user interface, making it an effective and easy solution for customers to implement.

DAI's commercialized version of the ASIST technology is working its way back to the Space Program. The technology was selected by Lockheed Martin Corporation as the in-house integration and test and mission operations system at its satellite production facility in Denver for the Origins Spectral Interpretation, Resource Identification and Security (OSIRIS) mission, which is scheduled to launch in 2011 to survey an asteroid and provide the first return of asteroid surface material samples to Earth, and for the Mars Atmosphere and Volatile Evolution (MAVEN) mission. The demonstrated cost savings is well over $3 million for the ground system element.

Northrop Grumman Corporation is also using designAmerica's NASA-developed technology as its basic rapid-development component for in-house project flight software development and avionics integration and test labs. This work benefits a range of satellites that the company is developing in-house, including for NASA missions such as the Lunar Crater Observation and Sensing Satellite (LCROSS). With ASIST at the core of its development suite, Northrop Grumman has demonstrated impressive gains in automated development and testing and streamlined satellite development, saving substantial mission dollars. ❖

Water-Based Coating Simplifies Circuit Board Manufacturing

Originating Technology/NASA Contribution

The Structures and Materials Division at Glenn Research Center is devoted to developing advanced, high-temperature materials and processes for future aerospace propulsion and power generation systems. The Polymers Branch falls under this division, and it is involved in the development of high-performance materials, including polymers for high-temperature polymer matrix composites; nanocomposites for both high- and low-temperature applications; durable aerogels; purification and functionalization of carbon nanotubes and their use in composites; computational modeling of materials and biological systems and processes; and developing polymer-derived molecular sensors. Essentially, this branch creates high-performance materials to reduce the weight and boost performance of components for space missions and aircraft engine components.

Under the leadership of chemical engineer, Dr. Michael Meador, the Polymers Branch boasts world-class laboratories, composite manufacturing facilities, testing stations, and some of the best scientists in the field.

Partnership

The Polymers Branch's extensive knowledge of polyimide chemistry and its expertise in the synthesis of ultraviolet-light-curable polyimides was the critical component that allowed Advanced Coatings International (ACI), of Akron, Ohio, to prototype the platform chemistry for a polyimide-based, waterborne, liquid photoimagable coating ideal for the manufacture of printed circuit boards.

Glenn and its partners in the Glenn Alliance for Technology Exchange (GATE), the Ohio Aerospace Institute (OAI), and Battelle's Great Lakes Industrial Technology Center (GLITeC) selected ACI as one of the winners of the GATE Partnership Award Program, a competition that targets small Ohio companies interested in enhancing their products and processes with NASA technologies.

ACI was one of 4 companies selected out of 38 applicants and received $50,000 to use toward product development, plus was eligible for an additional $50,000 to spend toward NASA assistance in developing their product. The company chose the additional NASA assistance, and Glenn set the company up with Meador to use his laboratories and polyimide chemistry expertise to develop the advanced water-based coating.

Product Outcome

Electronics manufacturers, constantly seeking ways to make their products smaller, lighter, and less expensive to produce, use advanced materials to meet ever-changing, demanding performance requirements. Often, they use polyimides as a substrate on which to form flexible circuit boards and create rigid/flexible hybrid circuits. The final step of this manufacturing process is to coat the circuit in an encapsulating, protective barrier, called a soldermask, usually a solvent-borne coating. These solvents tend to release harmful, volatile organic compounds (VOCs) into the atmosphere, both an environmental concern, and a safety hazard for employees exposed to the toxic fumes.

With Meador's assistance, ACI developed an advanced water-based coating that can be used in the manufacture of printed circuit boards in place of a traditional solvent-based formula. These water-based polyimide coatings are environmentally friendly and offer an improved level of worker safety. In fact, clean up requires just soap and water. Since they offer a safe, effective alternative to traditional solvent-based methods of electronics manufacturing, they are the preferred alternative for the environmentally conscious manufacturer.

In addition to improved safety and reduced environmental impact, these coatings have the potential to reduce manufacturing and operating costs. After working with a customer to evaluate this method, ACI now estimates that this new technology has the potential to save manufacturers of these devices operating costs of up to 25 percent, a figure that is sure to attract attention.

Additionally, the technology improves resolution, enabling manufacturers to create the smaller physical features required by today's electronics market. The environmentally friendly coating is photographically imaged onto the circuit board, providing a clear, precise, permanent protective layering.

Leveraging NASA expertise, ACI has managed to create an environmentally friendly, safe, and affordable process for manufacturing key components in a highly competitive, performance-driven market. ❖

Devoted to developing advanced, high-temperature materials and processes, Glenn Research Center's Structures and Materials Division works to create enabling technologies for future aerospace propulsion and power generation systems, such as the J-2X engine, planned to power the new crew launch vehicle's upper stage and the Earth-departure stage of the cargo launch vehicle.

Software Schedules Missions, Aids Project Management

Originating Technology/NASA Contribution

NASA missions require advanced planning, scheduling, and management, and the Space Agency has worked extensively to develop the programs and software suites necessary to facilitate these complex missions. These enormously intricate undertakings have hundreds of active components that need constant management and monitoring. It is no surprise, then, that the software developed for these tasks is often applicable in other high-stress, complex environments, like in government or industrial settings.

NASA work over the past few years has resulted in a handful of new scheduling, knowledge-management, and research tools developed under contract with one of NASA's partners. These tools have the unique responsibility of supporting NASA missions, but they are also finding uses outside of the Space Program.

Partnership

Knowledge-Based Systems Inc. (KBSI), of College Station, Texas, has worked with NASA on a handful of long-term **Small Business Innovation Research (SBIR)** contracts, which have ultimately allowed the company to develop several advanced technology solutions.

The first NASA SBIR contract was entered into with Johnson Space Center to create the Knowledge Aided Mission Planning System, or KAMPS, software for modeling and analyzing mission planning activities, simulating behavior and, using a unique constraint propagation mechanism, updating plans with each change in mission planning activities.

This successful NASA project led to another SBIR contract, this time with Kennedy Space Center, through which KBSI created another software solution, this one named the Optimization Modeling Assistant, or OMA, a set of tools to enable project managers to formulate optimization models—intelligent decisions made using

a ranked system of prioritized choices. The various decision-making programs developed by KBSI under these research projects with NASA have spun off into U.S. Air Force applications, have entered the commercial sector under the name of WorkSim, and have also been reintegrated into NASA programs where they have been used to model shuttle flows.

A third SBIR, again with Kennedy, resulted in the development of yet another software, this one a Web-based product, called the Portfolio Analysis Tool, or PAT, which allows teams of managers on the same project to

make strategic investment decisions. This tool is currently in use by NASA, the U.S. Air Force, and the U.S. Army.

Yet another SBIR with Kennedy allowed KBSI to create the Range Process Simulation Tool (RPST), which is now in use by the U.S. Army Black Hawk Fleet for operational performance.

The company also created the Toolkit for Enabling Adaptive Modeling and Simulation (TEAMS), through another Kennedy SBIR, which later, through a Phase III contract, evolved into the TEAMS+, a real-time operational analysis software that helps develop, maintain,

Knowledge-Based Systems Inc.'s AIOXFinder is an ideal search engine for sharing information in multisystem computer application environments.

and reconfigure operations analysis models and is the most recent commercialized product to have come out of KBSI's long history of NASA involvement, which includes years of working on NASA contracts since the company's founding in 1988.

United Space Alliance, one of the world's leading space operations companies, based near Johnson Space Center in Houston, is now using TEAMS for space shuttle ground processing to support operations analysis, planning, and scheduling. The program provides operations modeling and analysis for space transportation systems. According to Dr. Perakath Benjamin, KBSI vice president, the SBIR program has helped the company build and deploy advanced technology solutions that are benefiting NASA, the U.S. Department of Defense, and private industry.

Product Outcome

Three specific commercial products that KBSI developed trace their roots directly back to the work with NASA: WorkSim, Model Mosaic, and AIOXFinder.

WorkSim is a resource-constrained daily work-dispatching tool that generates optimized, daily schedules. It helps agencies streamline and speed their

Through partnering with NASA, Knowledge-Based Systems Inc. developed several advanced technology solutions that are aiding space missions and industry.

planning and scheduling and assists in the routine management of workflow and personnel, and its schedules can be generated in either Microsoft Excel or Project.

Model Mosaic is a knowledge-management work kit, much of which was developed through foundational work at Johnson Space Center. This software, while still under development, is based on the IDEF5 ontology description capture method that was developed by KBSI and is a government standard for object-oriented modeling. It allows users to extract the essential nature of concepts in the ontology domain and to document, in a structured manner, the behavior of entity relations in terms of the sanctioned inferences that can be made with them. The Model Mosaic toolkit includes an ontology management component that allows users to archive and, over time, to develop collections of ontologies that can be selectively analyzed, compared, pruned, and combined for use in novel systems development efforts. The Model Mosaic software also provides import and export capability to other popular ontology modeling languages like OWL and UML, and is plug-compatible with other OWL-based ontology editors.

AIOXFinder is an ontology-driven integration framework (ODIF), a type of search engine that is ideal for knowledge sharing and communication in large military enterprises or in multisystem computer application environments. The software refines a user's search according to that user's profile, or ontology, making it easier for researchers to weed out non-applicable hits in databases or through Internet searching. It is a document-centric, semantics-driven search engine, using text mining and natural language analysis to fine-tune search results. It is currently being tested by the U.S. Air Force at Cape Canaveral. ❖

Microsoft® and Excel® are registered trademarks of Microsoft Corporation.

The company has been able to apply the work it has done with NASA to several U.S. Air Force projects.

Software Analyzes Complex Systems in Real Time

Originating Technology/NASA Contribution

Expert system software programs, also known as knowledge-based systems, are computer programs that emulate the knowledge and analytical skills of one or more human experts, related to a specific subject. SHINE (Spacecraft Health Inference Engine) is one such program, a software inference engine (expert system) designed by NASA for the purpose of monitoring, analyzing, and diagnosing both real-time and non-real-time systems. It was developed to meet many of the Agency's demanding and rigorous artificial intelligence goals for current and future needs.

NASA developed the sophisticated and reusable software based on the experience and requirements of its Jet Propulsion Laboratory's (JPL) Artificial Intelligence Research Group in developing expert systems for space flight operations—specifically, the diagnosis of spacecraft health. It was designed to be efficient enough to operate in demanding real time and in limited hardware environments, and to be utilized by non-expert systems applications written in conventional programming languages.

The technology is currently used in several ongoing NASA applications, including the Mars Exploration Rovers and the Spacecraft Health Automatic Reasoning Pilot (SHARP) program for the diagnosis of telecommunication anomalies during the Neptune Voyager Encounter. It is also finding applications outside of the Space Agency.

Partnership

VIASPACE Inc., of Pasadena, California, a company dedicated to the commercialization of NASA and U.S. Department of Defense technologies, licensed the JPL-developed SHINE software from Caltech, which operates the laboratory for NASA. The company has an exclusive, worldwide license from Caltech to commercialize SHINE

technology for most major applications, including industrial, diagnostics/prognostics, sensor fusion, and homeland defense.

In January 2007, the company was awarded a prestigious NASA Space Act Award, which is given for significant scientific and technical contribution and provides recognition for those inventions and other scientific and technical contributions that have helped to achieve the Agency's aeronautical, technology transfer, and space goals.

Debbie Wolfenbarger, at JPL's Innovative Partnerships Program office, says of the partnership, "It is exciting to see a technology with such a rich history of being utilized in NASA missions make the leap to the commercial sector. Its potential in commercial applications such as diagnostics/prognostics and homeland defense applications will directly benefit the public and is an excellent demonstration of the positive effects resulting from licensing NASA technology."

Product Outcome

SHINE has been successfully applied to over a dozen applications both throughout NASA and within the commercial sector. It has significantly impacted operations cost, reliability, and safety in eight NASA deep space missions that include Voyager, Galileo, Magellan, and the Extreme Ultraviolet Explorer (EUVE).

SHINE has been used in the NASA Deep Space Network for diagnosing anomalies. It operates in a real-time environment, interfacing with in excess of 40 different legacy systems generating decisions and recommendations for control of the telecommunication systems. It has been employed by Ames Research Center as part of the Exploration Technology Development Program for onboard monitoring of flight system hardware and software with recovery. Johnson Space Center's Space Operations Management Office has used the software for technology transfer and experiments on robotics for crewed space flight. Marshall Space Flight Center has

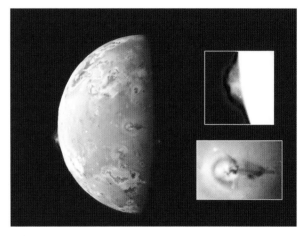

Two sulfurous eruptions are visible on Jupiter's volcanic moon Io in this color composite image from the robotic Galileo spacecraft that orbited Jupiter from 1995 to 2003. SHINE significantly decreased the operations cost for this mission, while increasing both reliability and safety.

used SHINE to conduct tests and evaluation experiments to support fault monitoring and abort system design for the Ares I.

JPL has also used the software both on its Engineering Analysis Subsystem Environment program, to operate a large number of spacecraft simultaneously through shared resources and automation, and in its EUVE, a NASA explorer-class satellite mission which employed SHINE to accomplish a labor reduction system, from three to one shift reductions through the use of artificial intelligence. On the Galileo mission, en route to Jupiter and its moons, NASA used SHINE for diagnosing problems in the Power and Pyro Subsystem. The Magellan mission, en route to Venus, benefited from SHINE's diagnosis of several telecommunication anomalies in the telecom subsystem.

As part of the Spacecraft Health Automatic Reasoning Pilot program, SHINE provided diagnosis of telecommunication anomalies during the Neptune Voyager encounter. This application prevented the possible loss of

SHINE has been applied to many NASA missions, including the Mars Exploration Rovers program.

the entire mission for detecting a failing transponder long before it was possible for a human to notice and just 12 hours prior to the encounter.

As part of JPL's X-33 Avionics Flight Experiment, SHINE was used as a Vehicle Health Manager.

It has also been used for several non-NASA aeronautics projects, including Lockheed Martin's Joint Strike Fighter, which involved proprietary efforts using SHINE in conjunction with the JPL-developed Beacon-based Exception Analysis for Multimissions (BEAM) software for fault diagnosis. SHINE was also successfully applied to anomaly detection on the F-18 aircraft. It was delivered as part of the BEAM anomaly detection system and ran on flight hardware, where it performed real-time mode identification on two F-18 aircraft engines and was tested in the F-18 hardware-in-loop (Iron Bird) simulator and during actual flight tests.

The Mars Exploration Rover mission is part of NASA's Mars Exploration Program, a long-term effort of robotic exploration of the Red Planet.

The U.S. Navy has employed SHINE as part of its Naval Sea Systems Command Integrated Vehicle Health Management program, for its capability to rapidly determine combat readiness in crisis situations and prognostics of network health coupled with remediation.

In addition, it has been applied to medical applications for Welch Allyn Inc., Johnson & Johnson, ViaLogy LLC, and VIASPACE. The company has also added new software capabilities that significantly enhance the functionality, flexibility, and ease of use of the SHINE system. The VIASPACE Knowledge Base Editor enables users to accelerate the development of SHINE-based applications and demonstrate capabilities and features for potential new markets and applications. ❖

Wireless Sensor Network Handles Image Data

Originating Technology/NASA Contribution

To relay data from remote locations for NASA's Earth sciences research, Goddard Space Flight Center contributed to the development of "microservers" (wireless sensor network nodes), which are now used commercially as a quick and affordable means to capture and distribute geographical information, including rich sets of aerial and street-level imagery.

NASA began this work out of a necessity for real-time recovery of remote sensor data. These microservers work much like a wireless office network, relaying information between devices. The key difference, however, is that instead of linking workstations within one office, the interconnected microservers operate miles away from one another. This attribute traces back to the technology's original use: The microservers were originally designed for seismology on remote glaciers and ice streams in Alaska, Greenland, and Antarctica—acquiring, storing, and relaying data wirelessly between ground sensors.

The microservers boast three key attributes. First, a researcher in the field can establish a "managed network" of microservers and rapidly see the data streams (recovered wirelessly) on a field computer. This rapid feedback permits the researcher to reconfigure the network for different purposes over the course of a field campaign. Second, through careful power management, the microservers can dwell unsupervised in the field for up to 2 years, collecting tremendous amounts of data at a research location. The third attribute is the exciting potential to deploy a microserver network that works in synchrony with robotic explorers (e.g., providing ground truth validation for satellites, supporting rovers as they traverse the local environment).

Managed networks of remote microservers that relay data unsupervised for up to 2 years can drastically reduce the costs of field instrumentation and data recovery.

Partnership

NASA did not create these microservers alone. Vexcel Corporation, of Boulder, Colorado, received **Small Business Technology Transfer (STTR)** funding through Goddard to develop the wireless sensor network technology that now aids in the high-speed handling of image data. This technology has uses in both the commercial sector, where it is used to relay satellite imagery to the desktop, and in the government sector, where NASA is finding continued use in terrestrial and interplanetary studies.

The STTR program is a three-phase program that reserves a specific percentage of Federal research and development funding for award to small businesses in partnership with nonprofit research institutions to move ideas from the laboratory to the marketplace, to foster high-tech economic development, and to address the technological needs of the Federal government. One of the few companies selected for this highly competitive program, Vexcel, a recognized world leader in the fields of photogrammetry, imagery, and remote sensing technologies, worked in partnership with Goddard and Pennsylvania State University to develop the new sensor network technology.

The work commenced in 2002, with an initial Phase I STTR award, the start-up phase for exploration

Photo - John McColgan BLM Alaska Fire Service

Firefighters could use the microserver technology to rapidly locate and track the spread of forest fires.

of the scientific, technical, and commercial feasibility of the technology. The work proved promising, and the company continued with a Phase II award in 2004. This second phase expands the work done under the first agreement, signals the initiation of the research and development phase, and is when the developer begins to consider commercialization potential. Vexcel excelled again in this phase and moved on to the highly prized Phase III, in which the company was able to bring the technology to market.

Soon after completing this long development process, Vexcel's new technology caught the eye of the Microsoft Corporation, which acquired the company in spring of 2006.

Vexcel's expertise in imaging technologies is now being applied to Microsoft's Virtual Earth business unit, incorporating the NASA-derived technology into its Microsoft Live Search, Virtual Earth service.

Product Outcome

Microsoft's goal in using the technology is to bring real-time imagery and other types of searchable data to the desktop. Using Vexcel's microserver technology, Microsoft hopes to add real-time information to its Live Search geographic search engine with many "information telepresence" applications. For example, firefighters could rapidly locate and track the spread of forest fires, and air traffic controllers could divert trans-Pacific flights away from ash plumes emanating from erupting Aleutian volcanoes.

Microsoft will use the microservers to gather real-time geospatial data (i.e., information connected to specific geographical locations) in order to provide users not only text data and maps from searches but also imagery including a bird's-eye view and three-dimensional pictures.

In addition to further internal development supported by Microsoft, the technology continues to be developed by Vexcel researchers under funding from NASA's Earth Science Technology Office's Advanced Information

On May 23, 2006, Jeff Williams, flight engineer of International Space Station Expedition 13, photographed this plume of ash produced by Cleveland Volcano, one of the most active of the volcanoes in the Aleutian Islands. The microserver technology could be used to divert air traffic from flying into plumes like this one, which spread as high as 6,000 meters (20,000 feet) above sea level.

Systems Technology program. Working under the acronym SEAMONSTER (South East Alaska Monitoring Network for Science, Telecommunications, Education, and Research), the project supports collaborative environmental science with near-real-time recovery of large volumes of environmental data. The initial geographic focus is at Alaska's Lemon Glacier and Lemon Creek watershed near Juneau, where the technology is being used to relay data about the glacier's effect on the hydrochemistry of Lemon Creek. Future expansion is planned in the Juneau Icefield and the coastal marine environment of the Alexander Archipelago.

This innovative sensor network technology may play a significant role in global climate research, as well as many other Earth science-related monitoring projects. In addition to these applications, Vexcel researchers are designing SEAMONSTER to be a powerful learning and teaching tool both through its construction and in planning for its future operation. ❖

Virtual Reality System Offers a Wide Perspective

Originating Technology/NASA Contribution

Robot Systems Technology Branch engineers at Johnson Space Center created the remotely controlled Robonaut for use as an additional "set of hands" in extravehicular activities (EVAs) and to allow exploration of environments that would be too dangerous or difficult for humans. One of the problems Robonaut developers encountered was that the robot's interface offered an extremely limited field of vision. Johnson robotics engineer, Darby Magruder, explained that the 40-degree field-of-view (FOV) in initial robotic prototypes provided very narrow tunnel vision, which posed difficulties for Robonaut operators trying to see the robot's surroundings. Because of the narrow FOV, NASA decided to reach out to the private sector for assistance. In addition to a wider FOV, NASA also desired higher resolution in a head-mounted display (HMD) with the added ability to capture and display video.

Partnership

Founded by former Johns Hopkins University optics researcher, Dr. Lawrence Brown, Sensics Inc., received a NASA **Small Business Innovation Research (SBIR)** program award in 2003, and began Phase I of their FOV interface for NASA's Robonaut, which also included development of a wide FOV camera. Baltimore-based Sensics created an HMD with a wide FOV, using a patented optical design to combine several tiled images into a high-resolution, three-dimensional panorama. A follow-on Phase II SBIR project culminated with Sensics delivering the piSight HMD to Johnson Space Center in February 2008, after delivering the prototype the year before.

The Sensics piSight HMD, essentially a high-tech viewfinder, offers an additional panoramic, high-resolution camera array that allows remote control and telepresence, immersing the HMD operator in the robot's workspace in real time. The operator sees through Robonaut's "eyes" and controls its movements, working safely from a distance. "Astronauts wearing a piSight will feel just like they are experiencing the world from Robonaut's perspective, due to the full field-of-view, which is an exciting advance for the NASA robotics program," says Brown.

Product Outcome

With the piSight, virtual surroundings appear in the viewfinder and respond to head movements. By coordinating head and joystick movements, the user flies or coasts up, down, forward, or sideways. Some versions of the piSight work with motion trackers, which eliminate the need for a joystick and, instead, follow the user's movement realistically.

When used with the piSight, InterSense Inc.'s SoniWing tracks user movements, employing technology similar to some video gaming systems. With the addition of the SoniWing, the user's entire body moves within the virtual environment, allowing greater freedom than systems restricted to head motion, and the environment responds realistically to the user's location and movements. Sensics also offers customized systems that track a user's eye movements, which help vision researchers learn which images or objects in a scene hold a user's interest, and also offers a video see-through solution for

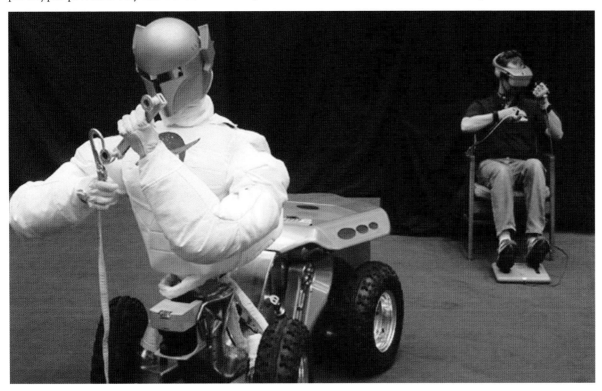

An operator in the background wears a head-mounted display that allows him to see through the Robonaut's eyes (cameras) with a wide near-human field-of-view.

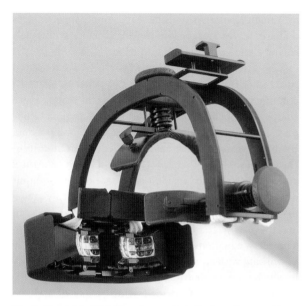

The lightweight piSight head-mounted display offers a wide field-of-view with multiple screens per eye to allow for more realistic visual experiences.

applications in which users need to see the virtual world and reality simultaneously.

The piSight HMD is light (weighing less than 2 pounds), stable, comfortable, and adjustable to an individual's head and eye shape, but Sensics still wants to make the technology more lightweight and efficient. Currently, piSight can be used with a variety of virtual reality software and platforms, including OpenSceneGraph on Linux and WorldViz products on Windows.

The piSight also offers high resolution "better than an HDTV display," according to Brown, with the NASA model displaying 12 screens per eye; lower-end models have as few as 3 screens per eye. Other commercially available HMDs usually offer fewer screens, fewer cameras, or poorer resolution, which leads to either a much smaller field of vision (i.e., no peripheral vision) or a poor-quality image that appears pixilated or stretched. With more screens per eye, however, each display can

offer a slightly different orientation, forming a concave arrangement that wraps around the eye, and projecting a different and more realistic perspective. In a higher-end model, the user can see in high resolution for 150 degrees, which is almost as wide as the natural range for humans. "The operator basically sees a field-of-view that has virtually no limitations," Brown notes.

The piSight camera array captures panoramic, high-resolution live video, which a user views while wearing the piSight HMD. Video is compressed, sent over a network, and then displayed inside the HMD. This interface may be able to support astronauts and engineers on Earth with various tasks like repairs and maintenance in high-risk environments. An astronaut in orbit, for example, could use the piSight remotely to control a robot on the surface of the Moon and examine the surrounding environment in real time. The system could also enhance safety on Earth by enabling remote operation of machinery in biohazard, defense, or medical environments, without requiring physical access to the site. In an emergency involving hazardous materials, a human operator could use the piSight system to remedy the situation from a safe location.

Sensics is finding success in the commercial sector, with sales of its products topping $1 million in 2007. Sensics is currently marketing the piSight to larger companies, since system prices start at about $27,500. Depending on customization and features, such as additional camera arrays for a wider field-of-vision, the units can cost up to $100,000.

Commercial applications for the piSight are only limited by customers' imaginations. "Customers have been selecting the piSight systems for a variety of simulation, training, design, and research applications," says Brown. An automobile designer uses the HMD to "sit" in the driver's seat of an automobile and explore the design in a realistic, immersive experience before production begins. Other companies are beginning to use the piSight for training purposes that involve learning specific

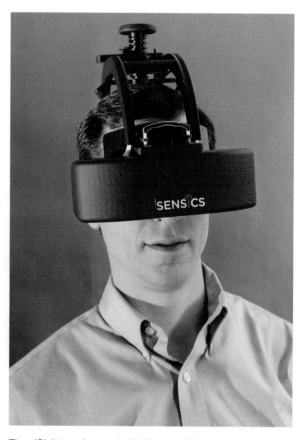

The piSight can be used with Linux or Windows virtual reality software, and can be controlled via motion trackers or joysticks.

tasks in situations, such as teaching a quarterback a specific play or showing a soldier how to operate a certain piece of equipment. Clearly, NASA will not be alone in reaping the benefits from virtual reality systems like the piSight. ❖

piSight™ is a trademark of Sensics Inc.
SoniWing™ is a trademark of InterSense Inc.
Linux® is a registered trademark of Linus Torvalds.
Windows® is a registered trademark of Microsoft Corporation.

Software Simulates Sight: Flat Panel Mura Detection

Originating Technology/NASA Contribution

In the increasingly sophisticated world of high-definition flat screen monitors and television screens, image clarity and the elimination of distortion are paramount concerns. As the devices that reproduce images become more and more sophisticated, so do the technologies that verify their accuracy. By simulating the manner in which a human eye perceives and interprets a visual stimulus, NASA scientists have found ways to automatically and accurately test new monitors and displays.

The Spatial Standard Observer (SSO) software metric, developed by Dr. Andrew B. Watson at Ames Research Center, measures visibility and defects in screens, displays, and interfaces. In the design of such a software tool, a central challenge is determining which aspects of visual function to include—while accuracy and generality are important, relative simplicity of the software module is also a key virtue. Based on data collected in ModelFest, a large cooperative multi-lab project hosted by the Optical Society of America, the SSO simulates a simplified model of human spatial vision, operating on a pair of images that are viewed at a specific viewing distance with pixels having a known relation to luminance.

The SSO measures the visibility of foveal spatial patterns, or the discriminability of two patterns, by incorporating only a few essential components of vision. These components include local contrast transformation, a contrast sensitivity function, local masking, and local pooling. By this construction, the SSO provides output in units of "just noticeable differences" (JND)—a unit of measure based on the assumed smallest difference of sensory input detectable by a human being. Herein is the truly amazing ability of the SSO—while conventional methods can manipulate images, the SSO models human perception. This set of equations actually defines a mathematical way of working with an image that accurately reflects the way in which the human eye and mind behold a stimulus.

The SSO is intended for a wide variety of applications, such as evaluating vision from unmanned aerial vehicles, measuring visibility of damage to aircraft and to the space shuttles, predicting outcomes of corrective laser eye surgery, inspecting displays during the manufacturing process, estimating the quality of compressed digital video, evaluating legibility of text, and predicting discriminability of icons or symbols in a graphical user interface.

Partnership

Radiant Imaging Inc., of Duvall, Washington, develops systems and software for testing and measuring color and light for lighting and display system designers, developers, manufacturers, and other users. To make use of the SSO's defect detection abilities in its work, the company licensed the software from Ames.

Already focused on the human perception of color and light in its line of colorimeters and photometers, Radiant Imaging sought this proven model as a means to design a system that reflected more perceptually meaningful criteria than gearing toward some scientific spectral characteristics. Ames provided training and consultation

The Spatial Standard Observer software measures visibility and defects in screens, displays, and interfaces.

to the engineers at Radiant Imaging regarding the SSO and related theory. The SSO software was then used to develop the TrueMURA Analysis Module. The module was incorporated into Radiant Imaging's proprietary ProMetric 9.1 system, originally created to provide automated defect detection analysis for flat panel display (FPD) systems. When used in conjunction with the ProMetric Series Imaging Colorimeters, the new software module provides a complete characterization and testing system for FPDs—especially LCD panels and displays—in research and development and production processes.

Automatic simulation of the human perception of display defects or blemishes—mura—is difficult, because human visual recognition of uniformity variations and other defects is dependent on many factors including brightness, color, and spatial relationships. Most available image analysis techniques are based on simple techniques which identify gross defects, but do not accurately discriminate JND. TrueMURA allows a grading of LCD mura in a way designed to mimic human observers, ignoring mura that cannot be seen by humans and ranking those likely to be distinguished by humans into different categories. TrueMURA is thus well-correlated to human perception of defects in brightness, color, and blemishes, and provides useful, quantifiable analysis of an FPD image.

Product Outcome

Radiant Imaging released the TrueMURA Analysis Module for ProMetric 9.1 in December 2007. This is the first commercial system available to provide advanced image analysis algorithms for computing JND. Augmenting the defect analysis functions offered by ProMetric 9.1, the TrueMURA Analysis Module has already been successfully integrated into thousands of systems in use worldwide.

"Working with NASA afforded Radiant Imaging access to cutting edge technology and expertise, and technology

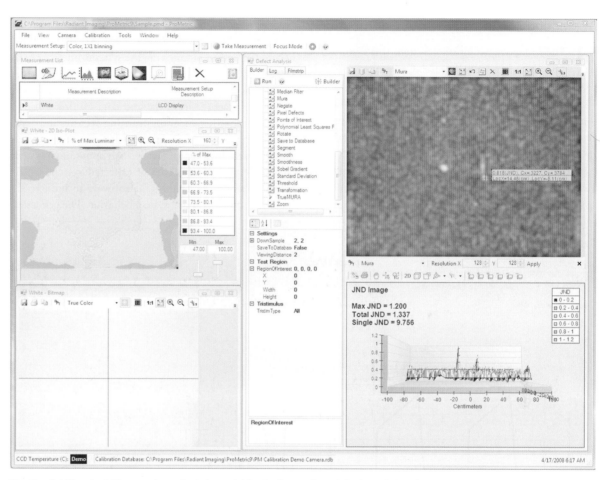

The Spatial Standard Observer is particularly useful for the inspection of displays during the manufacturing process.

transfer without commercial biases," said Dr. Hubert Kostal, of Radiant Imaging. "Together with PM Series Imaging Colorimeters, TrueMURA provides an accurate, repeatable, and objective means of assessing LCD display image quality."

The TrueMURA Analysis Module is now being integrated into Radiant Imaging's ProMetric Production Test Sequencer, used by display manufacturers to automate production testing of FPDs and projectors. By further automating defect detection, TrueMURA is expected to increase the speed and reduce the cost of product testing. While the primary applications for TrueMURA will continue to be display image quality, other applications for measuring visual uniformity are under investigation and anticipated in the near future. ❖

TrueMURA™ and PM Series™ are trademarks, and ProMetric® is a registered trademark of Radiant Imaging Inc.

Inductive System Monitors Tasks

Originating Technology/NASA Contribution

The Inductive Monitoring System (IMS) software developed at Ames Research Center uses artificial intelligence and data mining techniques to build system-monitoring knowledge bases from archived or simulated sensor data. This information is then used to detect unusual or anomalous behavior that may indicate an impending system failure.

Currently helping analyze data from systems that help fly and maintain the space shuttle and the International Space Station (ISS), the IMS has also been employed by NASA's hybrid combustion facility; an advanced rocket fuel test facility; the UH-60A RASCAL Black Hawk helicopter; and to monitor engine systems on an F/A-18 Hornet aircraft.

IMS uses techniques from the fields of model-based reasoning, machine learning, and data mining, though unlike some other machine learning techniques, IMS does not require examples of anomalous behavior, or failures, to learn. IMS automatically analyzes nominal system data to form general classes of expected system sensor values. This process enables the software to inductively learn and model nominal system behavior. The generated data classes are then used to build a monitoring knowledge base.

In real time, IMS performs monitoring functions: determining and displaying the degree of deviation from nominal performance. IMS trend analyses can detect conditions that may indicate a failure or required system maintenance.

The development of IMS was motivated by the difficulty of producing detailed diagnostic models of some system components due to complexity or unavailability of design information.

Successful applications have ranged from real-time monitoring of aircraft engine and control systems to anomaly detection in space shuttle and ISS data. IMS was used on shuttle missions STS-121, STS-115, and STS-116 to search the Wing Leading Edge Impact Detection System (WLEIDS) data for signs of possible damaging impacts during launch. It independently verified findings of the WLEIDS Mission Evaluation Room (MER) analysts and indicated additional points of interest that were subsequently investigated by the MER team.

In support of the Exploration Systems Mission Directorate, IMS is being deployed as an anomaly detection tool on ISS mission control consoles in the Johnson Space Center Mission Operations Directorate. IMS has been trained to detect faults in the ISS Control Moment Gyroscope (CMG) systems. In laboratory tests, it has already detected several minor anomalies in real-time CMG data. When tested on archived data, IMS was able to detect precursors of the CMG1 failure nearly 15 hours in advance of the actual failure event.

In the Aeronautics Research Mission Directorate, IMS successfully performed real-time engine health analysis. IMS was able to detect simulated failures and actual engine anomalies in an F/A-18 aircraft during the course of 25 test flights. IMS is also being used in collaboration with the Federal Aviation Administration's Aviation Safety Information Analysis and Sharing (ASIAS) system

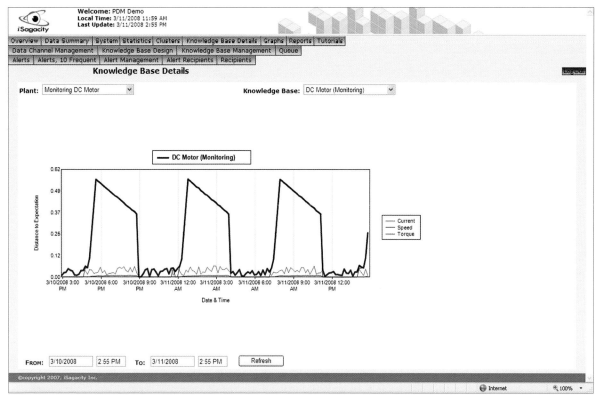

NASA partnered with industry to develop a new systems health monitoring software to increase safety and efficiency in complex industrial plants, such as power generation and water treatment facilities.

to analyze commercial airline data for anomalies that may indicate potential safety hazards.

Partnership

IMS can be used to monitor nearly any system with recurring behavior and appropriate data collection. This allows application to any number of system monitoring tasks.

iSagacity Inc., based out of Portland, Maine, executed a non-exclusive license with Ames for use of the IMS software. iSagacity was founded in 2001 as a provider of engineering software and consulting services to the process industries has a Web-based platform called Remote Manager, used for collecting data in real time, monitoring, and analyzing data for industrial process systems. Customer applications for the Remote Manager include water treatment plants, water heating and cooling in the process industry, oil refineries, public water distribution, and power generation plants.

In June 2006, iSagacity licensed IMS to complement the capabilities of its Remote Manager tool for monitoring industrial process systems.

The license is helping iSagacity to build early warning and diagnostics software systems for a variety of applications and industries. iSagacity CEO, Peter Millett, reports that the partnership with NASA has allowed the company to secure early stage development funding to develop and test its product.

Product Outcome

iSagacity has developed a new product, Process Data Miner, that uses IMS for a powerful new way of detecting trends in data that are early warning signs of process or equipment problems. Process Data Miner automatically scans data in real time and organizes it into a knowledge base consisting of "data signatures" or patterns that characterize operating behavior of the system. Process Data Miner automatically learns the patterns in data and organizes them for easy knowledge discovery.

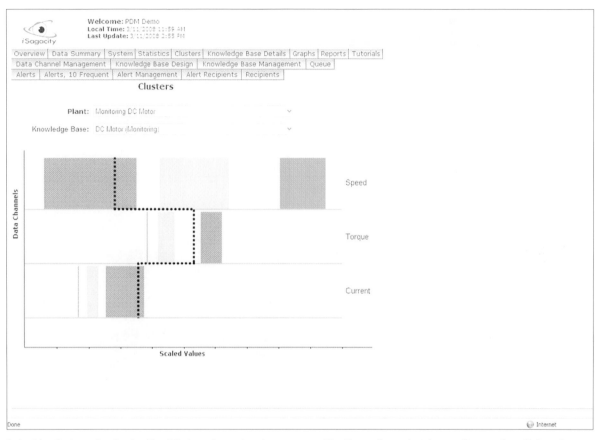

Industries that monitor the health of their equipment and processes with either online or batch sampling can benefit from the use of Process Data Miner.

iSagacity is pleased to count the NASA-developed software amongst its core technologies, and the IMS-based products will provide an elegant solution to the important problem of early warning and diagnostics for both industrial and commercial applications. Millett indicates that the use of the underlying IMS algorithms in its new product provides a very powerful way to rapidly build intelligence into equipment monitoring applications.

Other IMS applications under development include use in spacecraft and aircraft, and potential applications under consideration include telescope subsystems and uninhabited aerial vehicles. ❖

Remote Manager™ and Process Data Miner™ are trademarks of iSagacity Inc.

Mars Mapping Technology Brings Main Street to Life

Originating Technology/NASA Contribution

The Red Planet has long held a particular hold on the human psyche. From the Roman god of war to Orson Welles' infamous Halloween broadcast, our nearest planetary neighbor has been viewed with curiosity, suspicion, and awe. Pictures of Mars from 1965 to the present reveal familiar landscapes while also challenging our perceptions and revising our understanding of the processes at work in planets. Frequent discoveries have forced significant revisions to previous theories.

Although Mars shares many familiar features with Earth, such as mountains, plains, valleys, and polar ice, the conditions on Mars can vary wildly from those with which we are familiar. The apparently cold, rocky, and dusty wasteland seen through the eyes of spacecraft and Martian probes hints at a dynamic past of volcanic activity, cataclysmic meteors, and raging waters. New discoveries continue to revise our view of our next-door neighbor, and further exploration is now paving the way for a human sortie to the fourth stone from the Sun.

NASA's Mars Exploration Program, a long-term effort of robotic exploration, utilizes wide-angle stereo cameras mounted on NASA's twin robot geologists, the Mars Exploration Rovers (MERs), launched in 2003. The rovers, named "Spirit" and "Opportunity," celebrated 4 Earth years of exploration on January 3, 2008, and have sent back a wealth of information on the terrain and composition of the Martian surface. Their marathon performance has far outlasted the intended 90 days of operation, and the two intrepid explorers promise more images and data.

Partnership

earthmine inc., a street-level 3-D mapping company founded in 2006 and headquartered in Berkeley, California, licensed 3-D data-generation software and algorithms created by NASA's Jet Propulsion Laboratory (JPL) for the MERs from the California Institute of

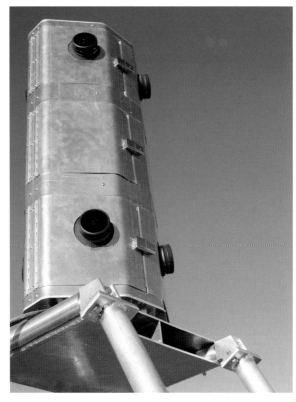

3-D data generation software and algorithms licensed by earthmine are currently utilized as a part of NASA's Mars Exploration Program. Wide-angle stereo cameras are mounted on NASA's twin robot geologists, the Mars Exploration Rovers, like those seen here on earthmine's vehicle-mounted array.

Technology (Caltech). The agreement with JPL and Caltech includes an exclusive and perpetual license for photogrammetric technology allowing for the creation of very dense and accurate 3-D data from stereo panoramic imagery.

earthmine combined the software and algorithms, originally used to create a 3-D representation of the local terrain to allow autonomous routing of the MERs through the Martian environment, with its unique capture

hardware and Web delivery technology. This system delivers 3-D data with unprecedented density, accuracy, and speed. earthmine utilizes the software and algorithms as a part of its processing pipeline, which automates the creation of high-quality, seamless panoramic imagery with pixel-for-pixel 3-D depth information from its image collection system.

"The JPL technology provides an unsurpassed level of accuracy and density in our 3-D data generation process," said earthmine co-founders John Ristevski and Anthony Fassero. "The problem of generating very dense and accurate 3-D data from wide-angle images is an extremely difficult one, which engineers and scientists at JPL have been working on solving for over a decade. Licensing this technology from JPL gives earthmine a significant advantage and will enable ongoing improvements in our technology."

Product Outcome

earthmine's core technology, called reality indexing, manages the flow of information from hardware, software, and workflow technologies. From streamlined data collection and data processing to Web-based dissemination, the system integrates the information to deliver accurate street-level geospatial data through a Web-based interface. Complete municipalities are collected through high-quality, 3-D panoramic images—including every road, alley, and freeway—to create a complete, consistent, and publicly accessible geospatial view of cities for official and commercial applications. As the cameras mounted on Spirit and Opportunity are designed to optimally record the Martian landscape, earthmine uses an automated vehicle-based camera array which is vertically oriented and has been optimized for the urban environment. This system captures entire metropolitan areas in a few weeks in 3-D, high-resolution panoramic images.

Within each immersive, high-resolution image, each pixel contains real-world latitude, longitude, and elevation information, and thus everything within an

The vehicle-mounted calibrated camera array is used to collect highly accurate and detailed data at the street level.

image can be accurately located, measured, or modeled using points, lines, or polygons. This data mine can be accessed through a Web-based interface where they can identify, view, and extract information as desired. The comprehensive field-of-view in the imaging system leaves no detail undocumented, allowing visualization of everything from overhead power lines and multistory buildings, to underlying road and curb features.

Current applications for the earthmine technology include a major chip manufacturer, mobile communications, satellite imagery, and asset management. earthmine technology has been incorporated by a 311 project in a major city, a county road project for a major metro area, a travel information company, an online real estate company, and a major architecture firm, among others. earthmine also plans to increase focus on solutions for cities and enterprises with strong geographic information

By visually connecting metadata to physical spaces, earthmine allows virtual inventories of our cities.

system (GIS) initiatives, the existing large and mature GIS system integrator channel that supports them, and the emerging market for Web-based mapping applications. For example, cities and major metropolitan areas can use earthmine resources in reviewing proposed new buildings, upgrading utilities, revising public transportation systems, and for municipal maintenance tasks. Additionally, due to the lightweight delivery system, earthmine has received interest from organizations in mobile navigation, which plan to employ the images as a tool to aid navigation in

a city through mobile telephones or in-car navigation systems.

"earthmine just might be the key to making the promise of the GeoWeb a reality," said Chris Shipley, executive producer of DEMO, an annual conference that serves as the launch vehicle for promising technologies and companies. "With its unique ability to put complex geospatial data in a visual context that anyone can understand, earthmine is enabling a new generation of mapping applications." ❖

Intelligent Memory Module Overcomes Harsh Environments

Originating Technology/NASA Contribution

Solar cells, integrated circuits, and sensors are essential to manned and unmanned space flight and exploration, but such systems are highly susceptible to damage from radiation. Especially problematic, the Van Allen radiation belts encircle Earth in concentric radioactive tori at distances from about 6,300 to 38,000 km, though the inner radiation belt can dip as low as 700 km, posing a severe hazard to craft and humans leaving Earth's atmosphere. To avoid this radiation, the International Space Station and space shuttles orbit at altitudes between 275 and 460 km, below the belts' range, and Apollo astronauts skirted the edge of the belts to minimize exposure, passing swiftly through thinner sections of the belts and thereby avoiding significant side effects. This radiation can, however, prove detrimental to improperly protected electronics on satellites that spend the majority of their service life in the harsh environment of the belts. Compact, high-performance electronics that can withstand extreme environmental and radiation stress are thus critical to future space missions.

Increasing miniaturization of electronics addresses the need for lighter weight in launch payloads, as launch costs put weight at a premium. Likewise, improved memory technologies have reduced size, cost, mass, power demand, and system complexity, and improved high-bandwidth communication to meet the data volume needs of the next-generation high-resolution sensors. This very miniaturization, however, has exacerbated system susceptibility to radiation, as the charge of ions may meet or exceed that of circuitry, overwhelming the circuit and disrupting operation of a satellite. The Hubble Space Telescope, for example, must turn off its sensors when passing through intense radiation to maintain reliable operation.

To address the need for improved data quality, additional capacity for raw and processed data, ever-increasing resolution, and radiation tolerance, NASA spurred the development of the Radiation Tolerant Intelligent Memory Stack (RTIMS). Suitable for both geostationary orbit (GEO) and low-Earth orbit (LEO) missions, RTIMS was developed with technology investments by Langley Research Center, Goddard Space Flight Center, ASRC Aerospace Corporation, Irvine Sensors Corporation, and 3D Plus USA Inc. Advances made by the RTIMS technology include:

- Significant reductions in the size and mass of memory arrays

- Simplified interface to a large synchronous dynamic random access memory (SDRAM) array with built-in logic for timing read, write, and refresh cycles

- Novel self-scrubbing and single event functional interrupt (SEFI) detection to improve radiation tolerance

- Radiation shielding and mitigation structure at the component level instead of system level to increase system reliability

- Added mission flexibility by operating in a triple modular redundancy (TMR) or in error detection and correction (EDAC) architecture, allowing RTIMS to be configured for the harshness of the expected environment

- In-flight reconfigurability using static random access memory (SRAM)-based field programmable gate array (FPGA) technology allows RTIMS to overcome both hardware and software errors that may be detected after launch, and adapt to changing mission conditions

Partnership

3D Plus USA Inc., of McKinney, Texas, licensed RTIMS from Langley for systems and methods to detect a failure event in FPGAs. One of the contributors to RTIMS—incorporating its circuit stacking technology—3D Plus has designed and manufactured many modules (memory-based, CPU, converters) for space applications, and has had its assembly process certified by NASA and the European Space Agency (ESA).

3D Plus is a leading company in 3-D electronics packaging in Europe and is recognized as a high-performance innovator in the design and manufacturing of miniaturized 3-D modules for active, passive, opto-electronics, and micro electro-mechanical systems (MEMS) and micro opto-electro-mechanical systems (MOEMS) components.

In partnership with Langley, 3D Plus developed the first high-density and fast access time memory module tolerant of space radiation effects. The project sought to develop and demonstrate an in-flight reconfigurable radiation tolerant stacked memory array based on state-of-the-art chip stacking, radiation shielding, and radiation mitigation technologies. The module can be directly connected to a processor or equivalent and operate like a radiation tolerant static random-access memory (SRAM) with a maximum capacity of 3Gb.

The first prototype intelligent and reconfigurable radiation tolerant memory modules were manufactured and tested by 3D Plus in April 2005. To manage total ionizing dose (TID) and single event effects (SEE) on electronics, specific hardware and patented software protections were developed. The resulting unit significantly reduces board design time as it requires no single event upset (SEU) management and no TID management. The result is a complete computing module—computing cores can be compiled into the onboard FPGA that can then support a distributed, reconfigurable computing architecture. This module decreases design complexity for space-based electronics requiring memory with its simple interface and internal radiation tolerance management (shielding, EDAC, overcurrent protections, and patented software protections).

Product Outcome

3D Plus introduced the commercial Intelligent Memory Module (IMM), intended for space applications such as commercial or scientific geostationary missions and deep space scientific exploration, and high-reliability

An artist's concept portraying NASA's Mars Science Laboratory, a mobile robot for investigating Mars' past or present ability to sustain microbial life. This picture is what the advanced rover, scheduled for launch in 2009, would look like in Martian terrain from a side aft angle. The mast, rising to about 2.1 meters (6.9 feet) above ground level, supports two remote-sensing instruments: the Mast Camera for stereo color viewing of surrounding terrain and material collected by the arm, and the ChemCam for analyzing the types of atoms in material that laser pulses have vaporized from rocks or soil targets up to about 9 meters (30 feet) away. The ChemCam incorporates the 3D Plus RTIMS.

computing in other radiation-intensive environments. Nuclear facilities, for instance, utilize automated or remote-controlled nuclear waste handlers that have significant computing and data collection tasks. Nuclear-powered craft, such as aircraft carriers, submarines, and proposed future spacecraft, would also be natural applications for this technology.

By using SRAM-based FPGAs, the IMM takes advantage of the inherent ability to change function and substantial data processing performance. The IMM provides a 3Gb memory capacity, including a TMR memory controller for multiple error correction. This controller has been applied to the XQ2V1000 SRAM-based FPGA from the Xilinx Virtex-II family. The modular architecture, SRAM-based FPGA re-programmability, and the large amount of SDRAM allow the implementation of custom applications, such as image data processing, high-speed network switching/monitoring, and other embedded applications.

3D Plus is collaborating with several partners on additional developments, including two applications that will see RTIMS return to space. EADS Astrium, Europe's leading satellite system specialist, is designing RTIMS into a satellite and evaluating it for other applications. Of particular interest, CNES (the French space agency) incorporated RTIMS into the ChemCam instrument for NASA's Mars Science Laboratory Rover mission, to be launched in 2009 and expected to reach Mars in 2010. ❖

Virtex® is a registered trademark of Xilinx Inc.

Integrated Circuit Chip Improves Network Efficiency

Originating Technology/NASA Contribution

Prior to 1999 and the development of SpaceWire, a standard for high-speed links for computer networks managed by the European Space Agency (ESA), there was no high-speed communications protocol for flight electronics. Onboard computers, processing units, and other electronics had to be designed for individual projects and then redesigned for subsequent projects, which increased development periods, costs, and risks. After adopting the SpaceWire protocol in 2000, NASA implemented the standard on the Swift mission, a gamma ray burst-alert telescope launched in November 2004. Scientists and developers on the James Webb Space Telescope further developed the network version of SpaceWire.

In essence, SpaceWire enables more science missions at a lower cost, because it provides a standard interface between flight electronics components; new systems need not be custom built to accommodate individual missions, so electronics can be reused. New protocols are helping to standardize higher layers of computer communication.

Goddard Space Flight Center improved on the ESA-developed SpaceWire by enabling standard protocols, which included defining quality of service and supporting plug-and-play capabilities. Goddard upgraded SpaceWire to make the routers more efficient and reliable, with features including redundant cables, simultaneous discrete broadcast pulses, prevention of network blockage, and improved verification. Redundant cables simplify management because the user does not need to worry about which connection is available, and simultaneous broadcast signals allow multiple users to broadcast low-latency side-band signal pulses across the network using the same resources for data communication. Additional features have been added to the SpaceWire switch to prevent network blockage so that more robust networks can be designed. Goddard's verification environment for the link-and-switch implementation continuously randomizes

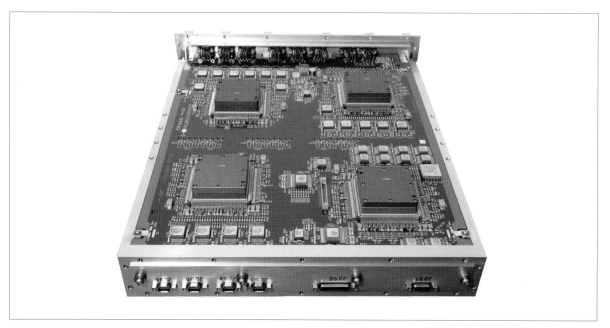

Goddard's SpaceWire offers the flexibility to meet a wide range of applications; it offers variable link rates, lower packet overhead, and provides the ability to multiplex discrete signals onto the same cable.

and tests different parts, constantly anticipating situations, which helps improve communications reliability. It has been tested in many different implementations for compatibility.

According to Goddard's Glenn Rakow, SpaceWire development lead, Goddard's SpaceWire implementation offers many advantages over other protocols. "It's very basic and simple, which translates to reliable implementations and flexibility for system engineers to meet a wide range of applications," said Rakow. Unlike Ethernet networks with pre-set link rates (10MHz, 100MHz, and 1GHz), SpaceWire helps save power and improve speed by offering variable link rates with a simple clock-recovery scheme. In addition, it provides lower packet overhead and a smaller implementation, and it provides the ability to multiplex traditional discrete signals onto the same cable.

Partnership

Headquartered in Rockville, Maryland, BAE Systems Inc. is a global company engaged in advanced electronics for aerospace systems. In order to bridge existing space electronics and Goddard's new SpaceWire design, the Agency formed a technology transfer partnership with the defense company in 2002 for the first-generation SpaceWire-based application-specific integrated circuit (ASIC) chip. Goddard integrated the first-generation SpaceWire ASIC on the geostationary operational environmental satellites R Series spacecraft, Advanced Baseline Imager, and NASA's Lunar Reconnaissance Orbiter single-board computer.

Goddard and BAE Systems collaborated again in 2006, when BAE Systems integrated the SpaceWire design into a new ASIC for the radiation-hardened RAD750 single

After adopting SpaceWire in 2000, NASA scientists and developers improved the network version of SpaceWire for the James Webb Space Telescope (shown here as a scale model).

Goddard and BAE Systems collaborated on a SpaceWire-based application-specific integrated circuit (ASIC) chip.

board. BAE Systems sought consultation from Goddard innovators to build a new ASIC incorporating the new SpaceWire implementation.

Goddard researchers and the Innovative Partnerships Program (IPP) team helped BAE Systems develop a benefits statement, statement of work, and a schedule for a Space Act Agreement. Because of its partnership, Goddard helped BAE Systems meet the tight ASIC design schedule. Rakow and the team at Goddard also provided technical support to BAE. According to Rakow, the collaboration between BAE Systems and Goddard is working well: "Industry has expertise that NASA doesn't have and vice versa, and that exchange will benefit the SpaceWire standard."

Product Outcome

The new BAE Systems ASICs are comprised entirely of reusable core elements, many of which are already flight-proven; they also incorporate a four-port SpaceWire router with two local ports, dual PCI bus interfaces, a microcontroller, 32KB of internal memory, and a memory controller for additional external memory use. The ASIC decreases the part count, overall communication system complexity, ongoing costs, and power requirements for the system's board while improving speed and reliability.

The SpaceWire-based ASIC appeals to commercial, military, and government aerospace customers because of its easy integration into onboard systems. Other industry SpaceWire implementations require multiple support chips rather than the single ASIC approach. Various aspects of SpaceWire illustrate its adaptability: Systems administrators can configure the ASIC according to the number of serial and local ports, while incorporating features from different systems. The ASIC is also available as a standalone chip for users to incorporate in their designs.

Goddard and BAE Systems continue to pursue ways to capitalize on their combined strengths. In 2008, new plans are underway to expand the current four-port SpaceWire ASIC to an eight-port implementation with an eye toward 2010 to rollout this design. ❖

NASA technologies aid industry. The benefits featured in this section:

- Revolutionize the welding industry
- Increase productivity in harsh environments
- Analyze rocks and minerals
- Strengthen welds
- Boost high-temperature performance
- Build better nanotubes

Novel Process Revolutionizes Welding Industry

Originating Technology/NASA Contribution

Deformation resistance welding (DRW) is a revolutionary welding process—a new technique for joining metals—in an industry that has not changed significantly in decades. Developed by the Energy and Chassis Division of the Detroit-based Delphi Corporation (a spinoff company formed by General Motors in 1999), DRW can reduce the cycle time and fabrication cost for a variety of structures using hollow members. Applications include automotive, aerospace, structural, and fluid handling applications.

As the name implies, DRW applies the heat and force of resistance welding, with tooling designed to create the necessary deformation. The process bonds metals atomically and creates solid-state joints through the heating and deformation of the mating surfaces, and as such, requires no additional filler materials. Metal tubes are joined to solid metal forms, sheet metal, and other tubes, creating nearly instantaneous, full-strength, leak-tight welds that can be used to build lean structural assemblies. The leak-tight joints are capable of holding fluids or gasses under pressure and heat, and can have strength exceeding that of the parent metals. DRW promises increases in performance and design flexibility while helping to cut cost, investment, and weight.

In early studies, DRW demonstrated improved quality over conventional welding methods and novel joining capabilities. These studies caught NASA's eye, and the Space Agency funded further development.

Partnership

NASA's Glenn Research Center, Delphi, and the Michigan Research Institute (MRI) entered into a research project to study the use of DRW in the construction and repair of stationary structures with multiple geometries and dissimilar materials, such as those NASA might use on the Moon or Mars. Delphi worked with the MRI (a not-for-profit organization created to speed the development

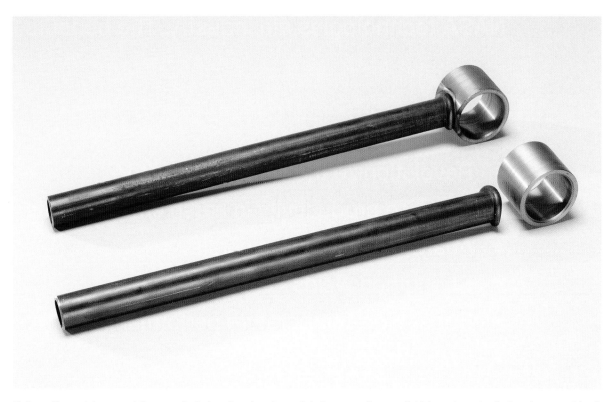

Deformation resistance welding uses include exhaust systems, tubular space frames, fluid-based mechanical systems, and load-bearing structural applications.

of emerging technologies) to obtain an initial $2.17 million in the form of two grants, which were used to help develop new weld joint design configurations, perfect existing welding techniques, and equip a laboratory with technicians to test DRW. NASA analyzed the test results to understand how DRW could be used to weld different types of metal on Earth and in space, with an eye toward eventually using DRW to weld structures on the Moon and Mars.

Initial testing proved promising, and in 2005, Delphi created SpaceForm Inc., also of Detroit, to commercialize the DRW technology. NASA, encouraged by the testing, provided additional funding for a joint research

partnership between Delphi, SpaceForm, and the Edison Welding Institute (EWI). Based in Columbus, Ohio, EWI is North America's leading engineering and technology organization dedicated to welding and materials joining. EWI's staff provides materials joining assistance, contract research, consulting services, and training to over 3,300 member company locations, representing world-class leaders in the aerospace, automotive, defense, energy, government, heavy manufacturing, medical, and electronics industries.

The EWI collaboration continued development of the DRW process and explored applications in ferrous and non-ferrous metals, dissimilar material joints, lean tubular

Deformation resistance welding's potential applications are numerous, though SpaceForm Inc. has targeted the automotive and non-automotive industries as ideal launch points for this state-of-the-art new technology.

structures, and concepts for future manufacturing cells. With EWI and NASA support, SpaceForm was able to advance DRW development toward production capabilities and narrow manufacturing parameters, making it a reliable, proven process. For these accomplishments, SpaceForm was awarded the Michigan Technology Leaders' "Corporate Partnership Award" for 2006.

"We're very pleased to have NASA's continued support," said Tim Forbes, director of commercialization and licensing at Delphi Technologies Inc., a subsidiary of Delphi Corporation. "The funded development projects with NASA have allowed us to gain a better understanding of the DRW joining process relative to a variety of materials, multiple joint configurations, joints with dissimilar materials, and the associated tooling and fixturing requirements. This work is supporting NASA objectives and is helping Delphi develop DRW for additional markets and customers."

Product Outcome

Current welding technologies are burdened by significant business and engineering challenges, including high costs of equipment and labor, heat-affected zones, limited automation, and inconsistent quality. DRW addresses each of those issues, while drastically reducing welding, manufacturing, and maintenance costs.

Manufacturers can expect lowered materials and capital costs and a significant reduction in welding cycle time. Additional advantages include localized heat application, and solid-state weld flexibility. The process can join dissimilar materials and shapes, is geometry-independent, and still automation friendly. Further, DRW is compliant (code case 2463) with the American Society of Mechanical Engineers and recognized for tube-to-tubesheet and heat exchanger manufacturing.

In addition to boasting a handful of manufacturing advantages, end-users will appreciate that the process eliminates weld leakage, that the weld is stronger than the parent metal, and that the method extends a product's service life. The process also eliminates tube thinning and porosity.

"The list of potential applications quickly grew from an initial chassis/suspension application," said Jayson D. Pankin, Delphi's new venture creation specialist and co-founder of SpaceForm Inc. "Other potential automotive applications, which could benefit from tube construction, like roof frames, cross-car beams, exhaust systems, and chassis module assemblies were immediate opportunities. The technology also has potential applications in medical devices, bridges, water heaters, plumbing, and more. Virtually anything that could benefit from tube welding."

DRW can reduce cycle time and cost in manufacturing an array of tubular structures. The technology can provide enhanced design flexibility in transportation and stationary applications, including motorcycles, recreational vehicles, bicycles, and wheelchairs, all while cutting cost investment in time and material. ❖

Deformation resistance welding drastically reduces welding, manufacturing, and maintenance costs.

Sensors Increase Productivity in Harsh Environments

Originating Technology/NASA Contribution

A team of scientists at Glenn Research Center, operating under the Aeronautics Research Mission Directorate's Aviation Safety and Fundamental Aeronautics programs, developed a series of technologies for testing aircraft engine combustion chambers. The team, led by electronics engineer Dr. Robert Okojie, designed a packaging technique and chip fabrication methods for creating silicon carbide (SiC) pressure sensors

to improve jet engine testing. According to Okojie, the team was "working to develop pressure sensors that would be used to more accurately measure pressure inside jet engine combustion chambers where the temperature is very high." Okojie's team also understood that "due to their temperature limitations, conventional pressure sensors are usually kept further away from the sensed environment. As a result, measurement accuracy is generally compromised. In addition, vital dynamic information could be lost due to frequency attenuation.

This new technology is meant to be inserted in close proximity with the sensed environment, thus eliminating these disadvantages."

SiC-based pressure sensors fabricated using these NASA technologies can operate for over 130 hours at 600 °C. These durable chips are applicable in engine ground testing and short-duration flight test instrumentation. Kathy Needham, director of the Technology Transfer and Partnership Office at Glenn, explains, "We have been spearheading the use of silicon carbide in sensors for some time now—the material withstands high temperatures and allows measurements to be taken closer to the source. As an added advantage, the new sensors are less complex than current, similar sensors, reducing the likelihood of performance failure, allowing them to be manufactured relatively inexpensively, and reducing system maintenance needs."

These factors also lend the technology to other uses, including commercial jet testing, deep well-drilling applications where pressure and temperature increase with drilling depth, and in automobile combustion chambers. As Okojie explains, "I see this technology being inserted in commercial jet engines and in deep wells while prospecting for oil. In commercial engine use, more accurate measurement of pressure would lead to improved engine safety by monitoring precursors of thermo-acoustic instability that leads to flame-out, efficient combustion of fuel and reduction of unwanted emission of hydrocarbons and nitrogen oxides. In deep well drilling, it would allow for longer term insertion into the wells, thereby significantly saving the cost of equipment maintenance and down time."

Partnership

To explore and develop these commercial uses, California's San Juan Capistrano-based Endevco Corporation, a division of the Meggitt PLC group, licensed three patents covering the high-temperature, harsh-environment silicon carbide pressure sensors from

Glenn Research Center licensed three patents, covering high-temperature, harsh-environment, silicon carbide pressure sensors, to Endevco Corporation.

Glenn. Collaboration between the two dates back to 2000, when, during combustion tests, Glenn used an Endevco silicon-based accelerometer as a benchmark device to validate its SiC accelerometer. The test results showed that the NASA device operated as well as the Endevco model, but the NASA-developed device had the added advantage of operating at much higher temperatures. This led to discussions between Endevco and Glenn about licensing opportunities to acquire Glenn's SiC pressure and accelerometer sensor fabrication and packaging technologies.

Endevco met with the Glenn researchers on numerous occasions during the following years. After witnessing the advantages of the NASA technology first-hand, the company licensed the three NASA patents for commercial development with the caveat that Okojie would continue to work on the project and help the company overcome any outstanding technical issues. Okojie agrees with the importance of this arrangement. "The transfer of this technology to the commercial sector," he says, "would extend the utilization of the products beyond the government (NASA, DOE, and DoD) to civil aviation, with derivative applications particularly in the efficient management of fuel combustion and the reduction of unwanted combustion byproducts."

Building on years of already successful collaboration, the partnership now steers these novel technologies toward their envisioned commercial applications. "Endevco is the ideal partner to bring this NASA technology to market, since the company already has a proven track record in the field and is now willing to commit to making this new technology accessible to industry," Needham elaborates. "The company," she continues, "is planning a new product line, enabling this high-performance NASA technology to achieve widespread use."

Product Outcome

Founded in 1947, Endevco supports its customers with a global network of manufacturing and research

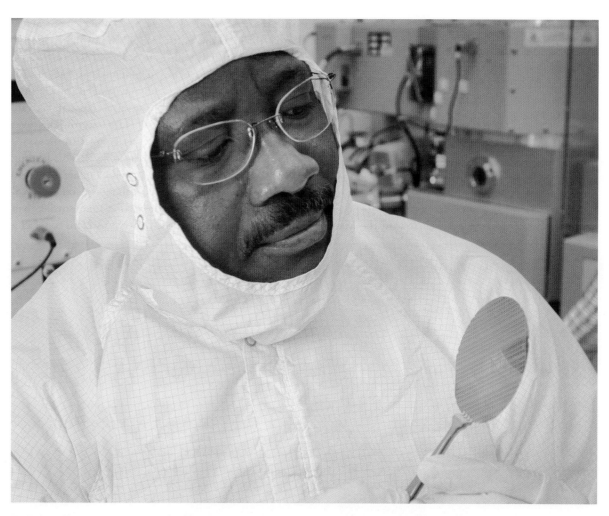

Dr. Robert Okojie, a researcher in the Sensors and Electronics Branch at Glenn, is a testament to the quality of a research environment that nurtures talent and ability at Glenn Research Center.

facilities, sales offices, and field engineers, providing trusted solutions for the world's most challenging measurement applications. In its more than 60 years of operation, Endevco has become a world leader in sensing solutions for demanding vibration, shock, and pressure applications. It was the company NASA trusted when it needed a benchmark device for its testing, and when the company saw that NASA had developed technologies that could improve the sensing field, it acted on the opportunity to license the cutting-edge know-how. NASA's SiC technology is Endevco's latest cutting-edge offering. ❖

Portable Device Analyzes Rocks and Minerals

Originating Technology/NASA Contribution

Building on the success of the two rover geologists that arrived on Mars in January 2004, NASA's next rover mission is being planned for travel to the Red Planet before the end of the decade. Twice as long and three times as heavy as the Mars Exploration Rovers, Spirit and Opportunity, the Mars Science Laboratory (MSL) will collect Martian soil and rock samples and analyze them for organic compounds and environmental conditions that could have supported microbial life.

MSL will be the first planetary mission to use precision landing techniques, steering itself toward the Martian surface similar to the way the space shuttle controls its entry through the Earth's upper atmosphere. In this way, the spacecraft will fly to a desired location above the surface of Mars before deploying its parachute for the final landing. As currently envisioned, in the final minutes before touchdown, the spacecraft will activate its parachute and retrorockets before lowering the rover package to the surface on a tether (similar to the way a sky crane helicopter moves a large object). This landing method will enable the rover to land in an area 20 to 40 kilometers (12 to 24 miles) long, about the size of a small crater or wide canyon and three to five times smaller than previous landing zones on Mars. NASA plans to select a landing site on the basis of highly detailed images sent to Earth by the Mars Reconnaissance Orbiter, in addition to data from earlier missions.

Like the twin rovers now on the surface of Mars, the MSL will have six wheels, and cameras mounted on a mast. Unlike the twin rovers, it will carry a laser for vaporizing a thin layer from the surface of a rock to analyze the elemental composition of the underlying materials, and will be able to collect rock and soil samples and distribute them to onboard test chambers for chemical analysis. Its design includes a suite of scientific instruments for identifying organic compounds such as proteins, amino acids, and other acids and bases that attach themselves to carbon backbones and are essential to life as we know it. MSL will also identify features such as atmospheric gasses that may be associated with biological activity.

Using these tools, the MSL will examine Martian rocks and soils in greater detail than ever before to determine the geologic processes that formed them; study the Martian atmosphere; and determine the distribution and circulation of water and carbon dioxide, whether frozen, liquid, or gaseous.

The rover will carry a radioisotope power system that generates electricity from the heat of plutonium's radioactive decay. This power source gives the mission an operating lifespan on Mars' surface of a full Martian year (687 Earth days) or more while also providing significantly greater mobility and operational flexibility, enhanced science payload capability, and exploration of a much larger range of latitudes and altitudes than was possible on previous missions to Mars.

Partnership

inXitu Inc., of Mountain View, California, a leader in portable X-ray diffraction/X-ray fluorescence (XRD/XRF) instrumentation, entered into a Phase II **Small Business Innovation Research (SBIR)** contract with Ames Research Center. The company specializes in developing technologies for the next generation of

The Mars Science Laboratory rover is larger and can travel farther than Spirit and Opportunity, NASA's two Mars Exploration Rovers that began exploring the Red Planet in early 2004.

scientific instruments for materials analysis. It rapidly evolved throughout the SBIR Phase II research, starting as a sole proprietorship and growing to a 10-employee corporation. Critical findings from the research were applied to the CheMin instrument (an instrument with chemical and mineral identification capabilities included in the analytical laboratory of MSL), enabling robust operation of its sample handling system.

During this SBIR Phase II research, inXitu designed and built an automated sample handling system for planetary XRD instruments that enables quality analysis of coarse-grained materials (XRD analyzes the crystalline arrangements of atoms or molecules in solids). The sample handling method developed by inXitu allows direct analysis of materials obtained from drills or rock crushers, eliminating the need for the extensive sample preparation typically required for XRD analysis. Sample loading and removal have been automated without complex mechanisms and with a minimum of moving parts.

inXitu's sample handling system could find a wide range of applications in research and industrial laboratories as a means to load powdered samples for analysis or process control. Potential industries include chemical, cement, inks, pharmaceutical, ceramics, and forensics. Additional applications include characterizing materials that cannot be ground to a fine size, such as explosives and research pharmaceuticals.

Product Outcome

inXitu's Terra product is the first truly portable XRD/XRF system designed specifically for rock and mineral analysis. XRD is the most definitive technique to accurately determine the mineralogical composition of rocks and soils. Phase identification is obtained by comparing the diffraction signature of a sample with a database of XRD mineral patterns. The addition of XRF informs on the nature of the chemical elements in the sample which allows easy screening during the phase identification process and can alleviate rare uncertainties.

inXitu's portable rock and mineral analysis device was tested at the Mars Analog site in Spitsbergen, Norway.

Terra is the result of over a decade of research and development for space exploration instrumentation. Using a low-power X-ray source and energy dispersive 2-D charge coupled device (CCD) detector, XRD and XRF data are obtained with no moving parts, providing an unparalleled robustness. The system provides fast identification capabilities thanks to its optimized geometry and high-sensitivity detector. With the company's patented sample handling system, only minimal sample preparation is required. Single minerals or simple mixtures can be identified after just a few minutes of integration.

Terra is built in a rugged, compact case for easy transport and operates autonomously via an embedded computer. In addition to a basic user interface on the instrument panel, a graphic user interface is available through a Wi-Fi link from any laptop or hand-held computer to control the instrument, preview live data, explore archive files, and download data.

Terra was first designed as a field geology/mineralogy instrument, but it can find a range of other applications where portability and fast analysis using XRD are required (mining, forensics, homeland security, etc.). Several Terra instruments are currently being used by NASA geologists to practice for the operational phase of the Mars Science Laboratory rover XRD instrument. ❖

NASA Design Strengthens Welds

Originating Technology/NASA Contribution

The Welding Institute (TWI), a nonprofit professional organization based out of the United Kingdom and devoted to the industry of joining materials, engineering, and allied technologies, developed a novel form of welding in the 1990s. Friction stir welding (FSW), the name under which it was patented, has been widely recognized as providing greatly improved weld properties over conventional fusion welds, and has been applied to manufacturing industries, including aircraft, marine, shipbuilding, including building decks for car ferries, trucking, railroading, large tank structure assembly, fuel tanks, radioactive waste container manufacturing, automotive hybrid aluminum, and the aerospace industry, where it is used to weld the aluminum external tank of the space shuttle.

FSW is a solid-state joining process—a combination of extruding and forging—ideal for use when the original metal characteristics must remain as unchanged as possible. During the FSW process, the pin of a cylindrical shouldered tool is slowly plunged into the joint between the two materials to be welded. The pin is then rotated at high speed, creating friction between the wear-resistant welding tool and the work piece. The resulting friction creates a plasticized shaft of material around the pin. As the pin moves forward in the joint, it "stirs," or crushes, the plasticized material, creating a forged bond, or weld.

Although the FSW process is more reliable and maintains higher material properties than conventional welding methods, two major drawbacks with the initial design impacted the efficacy of the process: the requirement for different-length pin tools when welding materials of varying thickness, and the reliance on a pin tool that left an exit, or "keyhole," at the end of the weld. The latter was a reliability concern, particularly when welding cylindrical items such as drums, pipes, and storage tanks, where the keyhole left by the retracted pin created a weakness in the weld and required an additional step to fill.

While exploring methods to improve the use of FSW in manufacturing, engineers at Marshall Space Flight Center (a licensee of TWI's technology) created new pin tool technologies, including an automatic retractable pin tool, to address the method's shortcomings. The tool uses a computer-controlled motor to automatically retract the pin into the shoulder of the tool at the end of the weld, preventing keyholes. The new technology addressed the limitations, and Marshall's innovative retractable pin tool has since contributed to customized FSW that has been proven to provide routinely reliable welds.

Since its invention, friction stir welding has received worldwide attention, and today, many companies around the world are using the technology in production, particularly for joining aluminum alloys.

Partnership

The NASA engineers patented their developments and sought commercial licensees for their new innovations. MTS Systems Corporation, of Eden Prairie, Minnesota, discovered the NASA-developed technology and then signed a co-exclusive license agreement to commercialize Marshall's auto-adjustable pin tool for a FSW patent in 2001. MTS is actively developing the FSW process, as well as new technologies to improve existing applications and to develop new ones. In addition, MTS has introduced the NASA technology to a wide variety of clients, in a wider variety of industries.

Product Outcome

MTS worked with the NASA technology and developed a flexible system that enables advanced FSW applications for high-strength structural alloys. The product also offers the added bonuses of being cost-competitive, efficient, and, most importantly, versatile. Customers include automotive, aerospace, and other industries.

The FSW system available from MTS offers users many advantages over conventional welding, including the ability to work with diverse materials; a wide range of alloys, including previously unweldable aluminums and high-temperature materials, while minimizing material distortion. This flexibility makes it adaptable to virtually any application.

The MTS system provides durable joints, with two to three times the fatigue resistance of traditional joining technologies like fusion welding or riveting, and has no keyholes. Since the pin is retracted automatically at the end of the welding process, and the hole is sealed, the technique produces consistent bonds.

MTS has supplied a FSW production system to an aluminum rim manufacturer in Norway. The Volvo XC 90 SUV rim will be manufactured using a cast aluminum face with a spun formed aluminum sheet (hybrid

The friction stir welding process is currently being used by several automotive companies and suppliers, including the manufacturing of wheel rims.

rim). The two pieces are friction stir welded by internal and external weld joints. Jim Freeman, MTS welding engineer, commented that "the customer was using an addition procedure to eliminate the keyhole; by introducing the retractable pin tool technology the cycle time was reduced by 70 percent."

FSW also lends itself to versatile welds, operating in all positions and able to create both straight welds and those requiring complex shapes. It even works with tapered-thickness weld joints, where the pin is able to maintain full penetration.

This technique, which does not require any consumables, is also advantageous, in that it does not create any environmental detriment, such as sparking, noise, or fumes. In addition, MTS's FSW process is also safer than conventional welding, since it does not create hazards such as toxic welding fumes, radiation, high voltage, liquid metal spatter, or arcing.

MTS is currently engaged in several research projects to expand the usefulness of FSW. While the company does not normally engage in research initiatives, these specific projects will assist customers in developing uses for the FSW technique.

First, MTS and an aircraft manufacturing customer developed a pin control mode utilizing the retractable pin tool. Using a surface sensor, the pin is located relative to the part surface using pin position control, while the shoulder is in force control. This control mode allowed the customer to weld a 0.012-inch-thick butt joint using the retractable pin tool.

The company also participated in a dual-use science and technology agreement with the U.S. Navy's Office of Naval Research to develop commercial and military applications for joining high-strength structural alloys. Program participants in addition to MTS include the University of South Carolina, General Dynamics Corporation's Bath Iron Works and Electric Boat, Oak Ridge National Laboratory, and the Naval Surface Warfare Center–Carderock Division.

In another research initiative to advance the uses of FSW in industrial applications, MTS, along with several other recognized industrial leaders (e.g., NASA, Boeing, Lockheed Martin, Spirit Aerosystems, General Motors) is a cooperative partner at the Center for Friction Stir Processing (CFSP). This is a multi-institutional National Science Foundation Industry/University Cooperative Research Center founded in August 2004. Methods like the retractable pin tool are a critical part of the CFSP technology development roadmap.

Having licensed the NASA technology and undertaken several initiatives to improve the commercial FSW technique, MTS is now in a unique position to contribute back to the Space Program, as its product has been selected by Marshall for use in welding upper-stage cryogenic hardware for the Constellation Program's Ares rockets. ❖

Polyimide Boosts High-Temperature Performance

Polyimide composites are introduced to the harsh environmental loads familiar to space launch propulsion systems in the combustion chamber of a rocket-based combined-cycle demonstrator.

Originating Technology/NASA Contribution

Polyimides are a class of polymers notable for chemical, wear, radiation, and temperature resistance, characteristics that have led to applications as diverse as aerospace engine housings and electronics packaging. Other applications include electronics, ranging from insulation for flexible cables to use as a high-temperature adhesive in the semiconductor industry. High-temperature polyimide carbon fiber composites are also used in non-loading structural components in aircrafts, weapon systems, and space vehicles. The appeal of polyimides is attributable to their unique combination of high-thermal stability, good chemical and solvent resistance, as well as excellent retention of mechanical properties at high temperature.

The polymerization of monomeric reactants (PMR) addition polyimide technology was developed in the mid-1970s at NASA's Lewis Research Center (renamed Glenn Research Center in 1999). This technology used an alcohol solution of polyimide monomers to make "prepreg," graphite or glass fiber bundles impregnated with polyimide resins, which could be thermally cured into composites with low voids, eliminating the difficulty of removing high-boiling solvents often used for condensation (step-polymerization) polyimides.

The initial PMR resin, known as PMR-15, is still commercially available and is used worldwide by the aerospace industry as the state-of-the-art resin for high-temperature polyimide composite applications; including engine bypass ducts, nozzle flaps, bushings, and bearings. PMR-15 can also be formualted into adhesives and coatings, and offers easy composite processing, excellent mechanical property retention for long-term use at temperatures up to 288 °C (550 °F), and is relatively inexpensive. As such, PMR-15 is widely regarded as a leading high-temperature polymer matrix composite for aircraft engine components.

However, PMR 15 is made from methylene dianiline (MDA), a known carcinogen and a liver toxin, and the Occupational Safety and Health Administration (OSHA) imposes strict regulations on the handling of MDA during the fabrication of PMR-15 composites. Recent concerns about the safety of workers involved in the manufacture and repair of PMR-15 components have led to the implementation of costly protective measures to limit worker exposure and ensure workplace safety.

Glenn researchers have continued to work on improving the properties and applicability of polyimides, with the ultimate goal of offering the aerospace, chemical, and automotive industries a lower toxicity alternative to PMR-15 that maintains similar processability, stability, and mechanical properties. The new polyimide developed by Dr. Kathy Chuang and Raymond Vannucci under NASA's Advanced Subsonic Technology (AST) program, named DMBZ-15, replaces MDA with a noncoplanar diamine, 2,2′-dimethylbenzidine (DMBZ).

The DMBZ-15 composition has a glass transition temperature of 414 °C (777 °F). This constitutes an increase in use temperature of 55 °C (100 °F) over the state-of-the-art PMR-15 composites and enables the development of fiber-reinforced polymer matrix composites with use temperatures as high as 343 °C (650 °F). DMBZ-15 graphite fiber reinforced composites exhibit an operational temperature range up to 335 °C (635 °F) and good thermo-oxidative stability in aircraft engine or missile environments.

Partnership

In 2002, Maverick Corporation, of Blue Ash, Ohio, licensed the DMBZ-15 technology.

In 2003, Chuang and Vannucci along with Maverick's Dr. Robert Gray and Eric Collins received R&D Magazine's 2003 "R&D 100" award recognizing DMBZ-15 among the best 100 new inventions of the year. In 2004, the DMBZ-15 technology received a NASA Space Act Award.

Product Outcome

DMBZ-15 bushings exhibit better wear resistance than state-of-the-art PMR-15. This ultrahigh-temperature material has a wide range of potential applications from aerospace (e.g., aircraft engine and airframe components, space transportation airframe and propulsion systems, and missiles) to bushings and bearings for non-aerospace applications (e.g., oil drilling, rolling mill). DMBZ-15 lightweight composites provide substantial weight savings and reduced machining costs compared to the same component made with more traditional metallic materials.

The DMBZ-15 resin can be made into lightweight bushings and polyimide/carbon fiber composite components for high-temperature applications.

Using solvent-assisted resin transfer molding, complex parts such as engine center vent tubes can be produced with braided reinforcement.

The DMBZ-15 polyimide has proven useful as a resin matrix with glass, quartz, and carbon fibers for lightweight, high-temperature composite applications similar to PMR-15 in aircraft engine components. Due to its higher temperature capability, it is especially suitable for use in missile applications, including fins, radon, and body components. Of particular interest to NASA, DMBZ-15 is well-suited to use as face sheets with honeycombs or thermal protection systems for reusable launch vehicles, which encounter elevated temperatures during launch and reentry. Other applications include use with chopped fibers to make bushings and bearings for engine or oil drilling components, and in high-temperature coating and ink applications. The light weight of DMBZ-15 polyimide composites invites use in secondary, non-load bearing aircraft engine components such as vent tubes, nozzle flaps and bushings, as well as for oil drilling components. Lightweight polymer composites also offer significant weight savings and subsequent improvements in fuel efficiency in aerospace propulsion and automotive applications. ❖

The "R&D 100" award-winning DMBZ-15 high-temperature polyimide.

NASA Innovation Builds Better Nanotubes

Originating Technology/NASA Contribution

One of the basic nanotechnology structures, the carbon nanotube, is a graphite sheet one atomic layer thick that is wrapped on itself to create an extraordinarily thin, strong tube. Although carbon nanotubes were discovered more than 15 years ago, their use has been limited due to the complex, dangerous, and expensive methods for their production. This unwieldy process has made widespread application of single-walled carbon nanotubes (SWCNTs) cost-prohibitive up until now.

Goddard Space Flight Center has made a major step forward in limiting these drawbacks. While traditional manufacturing methods use a metal catalyst to form the tubes, NASA researchers pinpointed this step as the cause of many of the drawbacks that were impeding development of SWCNTs. NASA researchers, under the direction of Goddard's Jeannette Benavides, discovered a simple, safe, and inexpensive method to create SWCNTs without the use of a metal catalyst. Benefits of this process include lowered manufacturing costs, a more robust product, and a simpler, safer process that produces a higher purity nanotube. NASA's SWCNT manufacturing process eliminates the costs associated with the use of metal catalysts, including the cost of product purification.

The removal of the catalyst not only reduces cost, but results in high-quality, very pure SWCNTs. Because NASA's process does not use a metal catalyst, no metal particles need to be removed from the final product. Eliminating the presence of metallic impurities results in the SWCNTs exhibiting higher degradation temperatures (650 °C rather than 500 °C) and eliminates damage to the SWCNTs by the purification process.

In addition to saving costs and creating a purer product, this new method also introduces features that make production simpler and safer. Unlike most current methods—which require expensive equipment (e.g., vacuum chamber), dangerous gasses, and extensive technical knowledge to operate—NASA's simple SWCNT manufacturing process needs only an arc welder, a helium purge, an ice-water bath, and basic processing experience to begin production. This simple method also offers an increase in quantity. Whereas traditional catalytic arc discharge methods produce an "as prepared" sample with a 30 to 50 percent SWCNT yield, NASA's method produces SWCNTs at an average yield of 70 percent.

Partnership

Nanotailor Inc., a nanomaterials company specializing in SWCNTs, based in Austin, Texas, licensed Goddard's unique SWCNT fabrication process with plans to make high-quality, low-cost SWCNTs available commercially.

"The nanotech industry is growing by more than 40 percent a year, but multi-walled carbon nanotubes have been the primary technology used. Single-walled technology just hasn't taken off because of the cost," notes Ramon Perales, president and chief executive officer of Nanotailor. "If we can get the cost down, we can be a step ahead and make higher quality nanotechnology more affordable." Other companies that have licensed the process include Idaho Space Materials Inc., a start-up in Boise, and American GFM Corporation, in Chesapeake, Virginia.

With a license agreement in place, Nanotailor built and tested a prototype based on Goddard's process. Device integrators and nanotechnology-based device companies are among Nanotailor's first customers, though the company hopes to cater to a wide variety of industries and research organizations.

Retired Goddard Space Flight Center researcher Jeannette Benavides prepares her award-winning, low-cost process for manufacturing high-purity, single-walled carbon nanotubes. This innovative manufacturing process uses helium arc welding to vaporize a carbon rod (anode), with the nanotubes forming in the soot deposited onto a water-cooled carbon cathode.

According to Dr. Reginald Parker, chief technical officer of Nanotailor, "Most industries using multi-walled tubes and technologies that require property improvement without a shift in weight will be able to benefit from this technology. A better product at a lower price will bring higher quality nanotechnology to biomaterials, advanced materials, space exploration, highway and building construction . . . the list seems endless as nanotubes have diverse and excellent properties."

Product Outcome

Nanotailor produces an optimized product based upon intellectual property licensed from NASA. Carbon nanotubes are being used in a wide variety of applications, and NASA's improved production method will increase their prevalence in the following areas:

- Medicine: SWCNTs offer the opportunity for improvements in many medical technologies, including implantable defibrillators (pacemakers); portable/field equipment; implantable biosensors; improved hearing aids; electrochemical analysis of biological materials; composites for long-lasting bone and joint implants; delivery of medicines and other treatments at the cellular level.

- Microelectronics: SWCNTs offer low resistance, low mass density, and high stability for improved microcircuits, nanowires, and transistors for miniature electronics; and consumer products, including pagers, cell phones, laptop and hand-held computers, toys, power tools, and automotive components.

- Scanning force/tunneling microscopy: Probing tips made with SWCNTs last longer and perform better than conventional silicon tips, improving materials science research and development; production quality control of semiconductor materials and data storage media; and evaluation of biological samples.

- Materials: SWCNTs do not affect a polymer's mechanical properties, allowing stress, transition, and

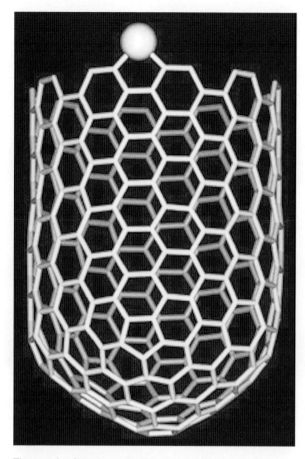

Thousands of times smaller than the average human hair, carbon nanotubes are extremely long and thin yet strong, making them a key nanotechnology structure.

thermal strain to be observed. SWCNTs also can be used to reinforce composites. Applications include dopants to create electrically conductive polymers; and easier monitoring of composites in critical applications (e.g., aircraft).

- Molecular containment: SWCNTs can be used to contain various elements, such as hydrogen for fuel cells and lithium boron hydrate for radiation shielding.

During September 2007, Nanotailor announced that its method for manufacturing high-quality carbon nanotubes was selected as a winner of the third annual Nanotech Briefs Nano 50 Awards. Jeannette Benavides, the inventor of the technology now retired from Goddard and serving as Nanotailor's director of research, was recognized at the NASA Tech Briefs National Nano Engineering Conference in Boston, November 14 and 15, 2007.

"We are very pleased that our technology received a Nano 50 award," commented Perales, "Nano 50 nominations are judged by a panel of nanotechnology experts of which the 50 technologies, products, and innovators with the highest scores are named winners. We are excited that our peers in the nanotechnology community believe that our technology has significantly impacted, or is expected to impact, the nanotechnology industry." ❖

Nanotech Briefs® is a registered trademark, and Nano 50™ is a trademark of ABP International Inc.

Inventor Jeannette Benavides was honored at the Nano 50 Awards ceremony.

Aeronautics and Space Activities

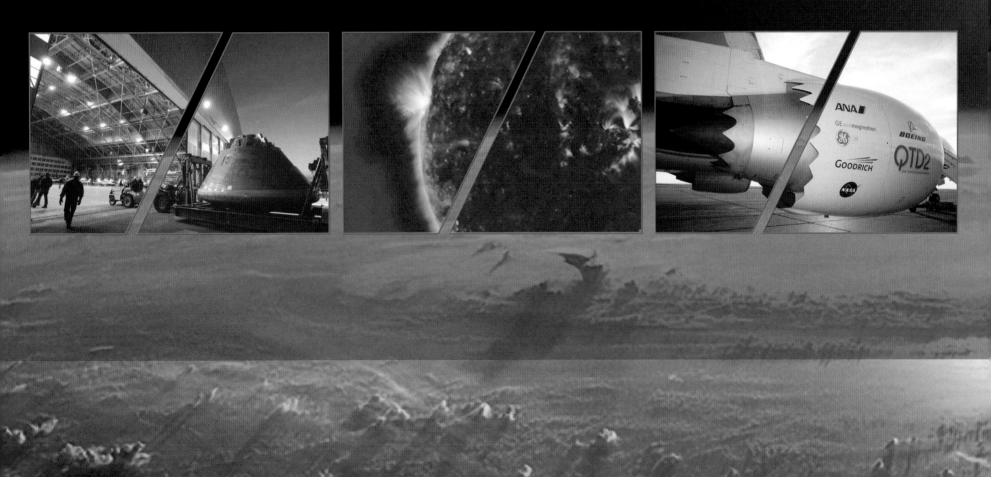

Pushing back boundaries in aeronautics and space exploration relies upon the ongoing research activities and operational support led by the four Mission Directorates: Exploration Systems, Space Operations, Science, and Aeronautics Research. These efforts are conducted at each of NASA's 10 field centers.

Aeronautics and Space Activities

Half a century after Congress enacted the National Aeronautics and Space Act of 1958, NASA is taking the next step in the Agency's proud tradition of exploration. As President Bush stated in 2004, when NASA's new focus on space exploration was announced, "We choose to explore space because doing so improves our lives, and lifts our national spirit."

The first generation of NASA reached the Moon with the Apollo program and unlocked the solar system with a rich legacy of robotic missions and satellites. Aeronautical vehicles pushed the air-space boundary and helped enable gains in aerospace and aviation. The space shuttle and International Space Station (ISS) mark the second generation of NASA's exploration journey. NASA has extended the sphere of human influence off the planet to 200 nautical miles into space, and now embarks on a new generational endeavor defined by its inspirational, bold, and practical spirit. With establishment of a lunar outpost, NASA will extend human influence to another planetary body, allowing exploration of the Moon, attainment of economic and scientific benefits, and the development of the ability to continue to extend the sphere of human influence to Mars and beyond.

This bold, new journey requires the strengths of the whole NASA team. The Exploration Systems Mission Directorate must develop the capabilities and technology that will enable sustained and affordable human and robotic exploration and ensure the health and performance of crews during long-duration space exploration, including robotic precursor missions, human transportation elements, and life support systems. The Space Operations Mission Directorate activities provide the communications, operational tests and evaluations, and mission operations competencies and assets. The Science Mission Directorate informs and is informed by our mutual solar system exploration activities. The Aeronautics Research Mission Directorate underpins the Agency's ability to create new vehicles and expand our operational regimes.

Exploration Systems Mission Directorate

NASA's Exploration Systems Mission Directorate (ESMD) develops capabilities and supporting research and technology that will make human and robotic exploration possible. It also makes sure that astronaut explorers are safe, healthy, and can perform their work during long-duration space exploration. In the near-term, ESMD does this by developing robotic precursor missions, human transportation elements, and life-support systems.

Lunar Outpost Plans Taking Shape

NASA's blueprints for an outpost on the Moon are shaping up. The Agency's Lunar Architecture Team has been hard at work, looking at concepts for habitation, rovers, and space suits.

NASA will return astronauts to the Moon by 2020, using the Ares and Orion spacecraft already under development. Astronauts will set up a lunar outpost—possibly near a South Pole site called Shackleton Crater—where they will conduct scientific research, as well as test technologies and techniques for possible exploration of Mars and other destinations.

Even though Shackleton Crater entices NASA scientists and engineers, they do not want to limit their options. To provide for maximum flexibility, NASA is designing hardware that would work at any number of

NASA's Lunar Architecture Team is looking at plans for space suits, habitats, and rovers; the Agency is considering small, pressurized rovers that would travel in pairs.

sites on the Moon. Data from the Lunar Reconnaissance Orbiter mission, a Moon-mapping mission set to launch in October 2008, might suggest that another lunar site would be best suited for the outpost.

NASA officials had been looking at having future moonwalkers bring smaller elements to the Moon and assemble them on site, but the Lunar Architecture Team found that sending larger modules ahead of time on a cargo lander would help the outpost get up and running more quickly. The team is also discussing the possibility of a mobile habitat module that would allow one module of the outpost to relocate to other lunar destinations as mission needs dictate.

NASA is also considering small, pressurized rovers that could be key to productive operations on the Moon's surface. Engineers envision rovers that would travel in pairs—two astronauts in each rover—and could be driven nearly 125 miles away from the outpost to conduct science or other activities. If one rover had mechanical problems, the astronauts could ride home in the other.

Astronauts inside the rovers would not need special clothing, because the pressurized rovers would have what is called a "shirt-sleeve environment." Space suits would be attached to the exterior of the rover. NASA's lunar architects are calling them "step in" space suits, because astronauts could crawl directly from the rovers into the suits to begin a moonwalk.

NASA is also looking to industry for proposals for a next-generation space suit. The Agency hopes to have a contractor onboard sometime this year.

NASA will spend the next several months communicating the work of the Lunar Architecture Team to potential partners—the aerospace community, industry, and international space agencies—to get valuable feedback that will help NASA further refine plans for the Moon outpost. The Agency's goal is to have finalized plans by 2012 to get "boots on the Moon" by 2020.

The concept for lunar rovers currently includes features such as all-direction crab steering with six wheels and a driver's perch that can pivot 360 degrees.

NASA's Newest Concept Vehicle Takes Off-Roading Out of This World

In a car commercial, it would sound odd: active suspension, six-wheel drive with independent steering for each wheel, no doors, no windows, no seats, and the only color available is gold.

But NASA's latest concept vehicle is meant to go way off-road, as in 240,000 miles from the nearest pavement, and drive on the Moon. Since NASA is working to send astronauts to the Moon by 2020 to set up a lunar outpost, it is going to need new wheels to help with scientific research and prepare for journeys to more distant destinations.

Built at Johnson Space Center, the new design is one concept for a future lunar truck. The vehicle provides an idea of what the transportation possibilities may be when astronauts start exploring the Moon. Other than a few basic requirements, the primary instruction given

to the designers was to throw away assumptions made on NASA's previous rovers and come up with new ideas.

"To be honest with you, it was scary when we started," said Lucien Junkin, a Johnson robotics engineer and the design lead for the prototype rover. "They tasked us last October to build the next-generation rover and challenge the conventional wisdom. The idea is that, in the future, NASA can put this side-by-side with alternate designs and start to pick their features."

One of the first standards to go was the traditional expectation that a vehicle should have four wheels. Mars rovers Spirit and Opportunity, still cruising around the Red Planet, have already proved the value of a couple of extra wheels. When one of Spirit's six wheels became inoperable, the rover had no problem rolling on using the remaining five.

With the number of wheels decided, the next question was how those wheels should turn. On a car, the front wheels turn a few inches in either direction, and both wheels point in the same direction. On this rover, all six wheels can pivot individually in any direction, regardless of where any other wheel points. To parallel park, a driver could pull up next to the parking place, turn all the wheels to the right, and slide right in.

Of course, astronauts will not have trouble finding a parking space on the Moon, but the feature, called crab steering, has advantages for a vehicle designed to drive into the craters of the Moon. If a slope is too steep to drive down safely, the vehicle could drive sideways instead—no backing up or three-point turns required. The all-wheels, all-ways steering also could come in handy when unloading and docking payloads or plugging into a habitat for recharging.

Introducing crab steering drove the concept in several other ways. If the rover's wheels turn to drive in a different direction, the driver needs to be able to do the same. The driver stands at the steering mechanism, because sitting in a space suit is not comfortable or practical. The astronaut's perch—steering mechanism, driver, and all—can pivot 360 degrees.

"The Apollo astronauts couldn't back up at all because they couldn't see where they were going in reverse," said Rob Ambrose, assistant chief of the Automation, Robotics, and Simulation Division at Johnson. "If you have a payload on the back or are plugging into something, it could be really important to keep your eyes directly on it."

The vehicle also can be the ultimate low-rider. It can lower its belly to the ground, making it easier for astronauts in space suits to climb on and off. Individual wheels or sections can be raised and lowered to keep the vehicle level when driving on uneven ground.

Some, all, or none of these features may be selected for the design of a rover that eventually goes to the Moon. Even though NASA's lunar architects currently envision pressurized rovers that would travel in pairs, with two astronauts in each rover, the new prototype vehicle is meant to provide ideas as those future designs are developed.

"This rover concept changed the whole paradigm," said Diane Hope, program element manager for NASA's Exploration Technology Development Program at Langley Research Center, which sponsored the vehicle's development. "It's not something I would have expected. It provides an alternative approach."

NASA Team Demonstrates Robot Technology for Moon Exploration

During the 3rd Space Exploration Conference, February 26-28, 2007, in Denver, NASA exhibited a robot rover equipped with a drill designed to find water and oxygen-rich soil on the Moon.

"Resources are the key to sustainable outposts on the Moon and Mars," said Bill Larson, deputy manager of the In Situ Resource Utilization (ISRU) project. "It's too expensive to bring everything from Earth. This is the first step toward understanding the potential for lunar resources and developing the knowledge needed to extract them economically."

The engineering challenge was daunting. A robot rover designed for prospecting within lunar craters has to operate in continual darkness at extremely cold temperatures with little power. The Moon has one-sixth the gravity of Earth, so a lightweight rover will have a difficult job resisting drilling forces and remaining stable. Lunar soil, known as regolith, is abrasive and compact, so if a drill strikes ice, it likely will have the consistency of concrete.

Meeting these challenges in one system took ingenuity and teamwork. Engineers demonstrated a drill capable of digging samples of regolith in Pittsburgh last December. The demonstration used a laser light camera to select a site for drilling, and then commanded the four-wheeled rover to lower the drill and collect 3-foot samples of soil and rock.

"These are tasks that have never been done and are really difficult to do on the Moon," said John Caruso,

A conceptual illustration shows a lunar robot rover equipped with a drill designed to find water and oxygen-rich soil.

demonstration integration lead for ISRU and Human Robotics Systems at Glenn Research Center.

In 2008, the team plans to equip the rover with ISRU's Regolith and Environment Science and Oxygen and Lunar Volatile Extraction experiment, known as RESOLVE. Led by engineers at Kennedy Space Center, the RESOLVE experiment package will add the ability to crush a regolith sample into small, uniform pieces and heat them.

The process will release gasses deposited on the Moon's surface during billions of years of exposure to the solar wind and bombardment by asteroids and comets. Hydrogen is used to draw oxygen out of iron oxides in the regolith to form water. The water then can be electrolyzed to split it back into pure hydrogen and oxygen, a process tested earlier this year by engineers at Johnson.

"We're taking hardware from two different technology programs within NASA and combining them to demonstrate a capability that might be used on the Moon," said Gerald Sanders, manager of the ISRU project. "And even if the exact technologies are not used on the Moon, the lessons learned and the relationships formed will influence the next generation of hardware."

Engineers participated in the ground-based rover concept demonstration from four NASA centers; the Canadian Space Agency; the Northern Centre for Advanced Technology, in Sudbury, Ontario; and Carnegie Mellon University's Robotics Institute, in Pittsburgh.

Carnegie Mellon was responsible for the robot's design and testing, and the Northern Centre for Advanced Technology built the drilling system. Glenn contributed the rover's power management system. NASA's Ames Research Center built a system that navigates the rover in the dark. The Canadian Space Agency funded a Neptec camera that builds 3-D images of terrain using laser light.

All the elements together represent a collaboration of the Human Robotic Systems and ISRU projects at Johnson, which are part of Langley's Exploration Technology Development Program.

In preparation for the next generation of exploration vehicles, NASA's Constellation Program has scheduled the first unmanned abort test for 2008.

NASA Readies Hardware for Test of Astronaut Escape System

Returning humans to the Moon by 2020 may seem like a distant goal, but NASA's Constellation Program already has scheduled the first test flight toward that goal to take place in less than a year.

The 90-second flight will not leave Earth's atmosphere, but it will be an important first step toward demonstrating how NASA intends to build safety into its next generation of spacecraft, including the Ares I and V rockets and the Orion crew capsule.

The first in a series of unmanned abort tests, known as Pad Abort-1 or PA-1, is scheduled for late 2008 at the U.S. Army's White Sands Missile Range in New Mexico.

The tests will help verify that NASA's newly developed spacecraft launch abort system can provide a safe escape route for astronauts in the Orion crew capsule in the event of a problem on the launch pad or during ascent into low-Earth orbit atop the Ares I rocket.

Orion is the Constellation Program's new crew exploration vehicle, set to carry as many as four crew members to lunar orbit and return its crew safely to Earth after missions to the Moon's surface. The 5-meter (16.5-foot)-wide, cone-shaped capsule also will provide transport services to the International Space Station (ISS) for as many as six crew members. Before launching to the Moon or to the ISS, however, system tests on Earth have to prove that the technologies work.

The pad abort test will simulate an emergency on the launch pad. Upon command from a nearby control center, a dummy Orion crew module—which would sit on top of a rocket for an actual launch—will be ejected directly from the launch pad by its rocket-propelled launch abort system to about 1 mile in altitude and nearly 1 mile downrange.

That is why engineers and technicians at Langley, Dryden Flight Research Center, and industry partners on the Orion Project are taking particular care to fabricate and equip the first flight test articles with extreme precision.

Engineers and technicians at Langley designed and fabricated the structural shell of the simulated crew module for the first pad abort test and now are conducting a series of ground checks on the structure. The crew module simulator accurately replicates the size, outer shape, and mass characteristics of the Orion crew module.

"The next step is to ship the completed crew module simulator to Dryden, where they will outfit it with the smarts—the computers, the electronics, the instrumentation—all the systems that need to work in conjunction with the structure," said Phil Brown, manager of the Langley Orion Flight Test Article Project.

After the instrumented dummy crew module is delivered to White Sands, it will be integrated with the PA-1 launch abort system flight test article, a vertical tower containing the escape rocket motor and a guiding rocket motor currently under construction at Orbital Sciences Corporation, in Dulles, Virginia. The combined crew module and launch abort system will be placed on the launch pad being constructed especially for the abort flight test series.

During the pad abort test sequence, the escape system's main abort motor will fire for several seconds, rapidly lifting the simulated crew module from the test launch pad, after which the escape system will detach, and three 116-foot-diameter parachutes will deploy to slow the module for landing.

The test will provide early data for design reviews to follow and will be followed by an ascent abort test in 2009 and a second pad abort test scheduled for 2010, both at White Sands. A parallel series of higher-altitude launch tests will commence at Kennedy in 2009.

"These flight tests will either confirm that our system works or help us identify and correct any defects that surface," said Greg Stover, manager of the Orion Launch Abort System Project Office, located at Langley. "Our goal is that on every manned mission, the launch abort system will be the most reliable system that we hopefully never have to use."

In addition to Langley, Dryden, and Kennedy, the Orion Project launch abort system team and the abort flight test team includes members from Johnson, Glenn, and Marshall Space Flight Center in Huntsville—as well as Orion Project prime contractor, Lockheed Martin Corporation, of Denver, and its subcontractor, Orbital Sciences.

The Orion Project Office, located at Johnson, is leading the development of the Orion spacecraft for the Constellation Program, which also includes the Ares I and Ares V launch vehicles, the Altair human lunar lander, and lunar surface systems to support sustained crew habitation.

Space Operations Mission Directorate

The Space Operations Mission Directorate provides NASA with leadership and management of the Agency's space operations related to human exploration in and beyond low-Earth orbit. Space Operations also oversees low-level requirements development, policy, and programmatic oversight. Current exploration activities in low-Earth orbit include the space shuttle and International Space Station (ISS) programs. The directorate is similarly responsible for Agency leadership and management of NASA space operations related to launch services, space transportation, and space communications in support of both human and robotic exploration programs. Its main challenges include: completing assembly of the ISS; utilizing, operating, and sustaining the ISS; commercial space launch acquisition; future space communications architecture; and transition from the space shuttle to future launch vehicles.

Dextre: Canadian Robotics for the International Space Station

Dextre is the third and final component of the Mobile Servicing System (MSS) developed by Canada for the ISS. The two-armed Special Purpose Dexterous Manipulator complements the mobile base and the robotic arm Canadarm2 already installed and operating on the station. These make the MSS a vital tool for external station maintenance. With advanced stabilization and handling capabilities, Dextre can perform delicate human-scale tasks such as removing and replacing small exterior components. Operated by crew members inside the station or by flight controllers on the ground, it also is equipped with lights, video equipment, a stowage platform, and three robotic tools.

The technology behind Dextre evolved from its famous predecessor, Canadarm2. Dextre is the world's first on-orbit servicing robot with an operational mission, and it lays the foundation for future satellite servicing and space exploration capabilities.

While one arm is used to anchor and stabilize the system, the other can perform fine manipulation tasks such as removing and replacing station components, opening and closing covers, and deploying or retracting mechanisms. Dextre can either be attached to the end of Canadarm2 or ride independently on the mobile base system. To grab objects, Dextre has special grippers with a built-in socket wrench, camera, and lights. The two pan/tilt cameras below its rotating torso provide operators with additional views of the work area.

Currently, astronauts execute many tasks that can only be performed during long, arduous, and potentially dangerous spacewalks. Delivery of this element increases crew safety and reduces the amount of time that astronauts must spend outside the station for routine maintenance. They should, therefore, have more time for scientific activities.

Some of the many tasks Dextre will perform include:

- Installing and removing small payloads such as batteries, power switching units, and computers
- Providing power to payloads
- Manipulating, installing, and removing scientific payloads

A typical task for Dextre would be to replace a depleted battery (100 kg) and engage all the connectors. This involves bolting and unbolting, as well as millimeter-level positioning accuracy for aligning and inserting the new battery.

This kind of task demands high precision and a gentle touch. To achieve this, Dextre has a unique technology: precise sensing of the forces and torque in its grip with automatic compensation to ensure the payload glides smoothly into its mounting fixture. Dextre can pivot at the waist, and its shoulders support two identical arms with seven offset joints that allow for great freedom of movement. The waist joint allows the operator to change the position of the tools, cameras, and temporary stowage on the lower body with respect to the arms on the upper body. Dextre is designed to move only one arm at a time for several reasons: to maintain stability, to harmonize activities with Canadarm on the shuttle and Canadarm2 on the station, and to minimize the possibility of self-collision.

At the end of each arm is an orbital replacement unit/tool change-out mechanism, or OTCM—parallel jaws that hold a payload or tool with a vice-like grip. Each OTCM has a retractable motorized socket wrench to turn bolts and mate or detach mechanisms, as well as a camera and lights for close-up viewing. A retractable umbilical connector can provide power, data, and video connection feed-through to payloads.

From a workstation aboard the ISS, astronauts can operate all the Mobile Servicing System components, namely Canadarm2, the mobile base, and Dextre. To prepare for operating each component, astronauts

The Canadarm2 aboard the International Space Station has multiple joints and is capable of maneuvering payloads as massive as 116,000 kilograms, equivalent to a fully loaded bus. Pictured above, astronaut Stephen Robinson rides Canadarm2 during the STS-114 mission of the Space Shuttle Discovery to the ISS in August 2005.

and cosmonauts undergo rigorous training at the Canadian Space Agency's Operations Engineering Training Facility at the John H. Chapman Space Centre, in Longueuil, Quebec.

Renowned for its expertise in space robotics, Canada's contribution to the International Space Station—a unique collaborative project with the United States, Japan, Russia, and several European nations—is the Mobile Servicing System. Combining two robotic elements and a mobile platform, they are designed to work together or independently. The first element, Canadarm2, whose technical name is the Space Station Remote Manipulator System, was delivered and installed by Canadian Space Agency astronaut Chris Hadfield in 2001. The mobile base system was added to the station in 2002. Dextre launched aboard Space Shuttle Endeavour flight STS-123.

Space Station Provides New 'Window' for International Polar Year

It has happened only three other times in history. But this time, the International Polar Year will have unprecedented access to an out-of-this-world platform—210 miles up in space.

The International Polar Year is a collaborative effort to study the Arctic and the Antarctic from March 2007 to March 2009. Crew members on the ISS are supporting the scientific program by taking new snapshots of Earth's polar regions—the areas of the globe surrounding the North and South poles—from the unique vantage point of space.

It has been 50 years since the last event like this, with previous polar years observed in 1882, 1932, and 1957. This year will have the largest number of participants, with thousands of scientists from more than 60 countries examining the polar regions in a wide range of topics, including surface and atmospheric temperatures, changes in snow and ice, and the shrinkage of glaciers. They also hope to dip into unexplored areas.

Because the ISS provides a unique venue for observing polar phenomena, NASA has invited scientists participating in the polar program to submit requests for relevant imagery to be photographed from space.

As part of the Crew Earth Observations experiment, space station astronauts photograph designated sites and dynamic events on the Earth's surface using digital cameras equipped with a variety of lenses. Depending on the station's position and weather in the target regions, astronauts can collect high-resolution digital photos of a specific location or lower-resolution photos that cover very large areas.

During the timeframe of the International Polar Year, polar observations are a scientific focus for the Crew Earth Observations experiment (CEO-IPY).

Previous space station crew members have photographed such phenomena as auroras, polar mesospheric clouds and patterns, and calving of sea ice. Although the station does not cross the poles, astronauts can look toward the poles to document these phenomena.

"Polar regions are rich in phenomena obtuse to our daily temperate-region lives and naturally draw one's scientific attention, whether on or off of Earth," said NASA astronaut Don Pettit, the leader of the effort to link polar scientists with the space station. "I want IPY scientists on Earth to have access to the space station perspective, where observations can be made on the length scale of half a continent and will complement observations made on Earth or by higher orbiting satellites."

The Crew Earth Observations imagery Web site for CEO-IPY, **<http://eol.jsc.nasa.gov/ipy/>**, provides an online form that allows polar region investigators to interact with Earth observation scientists and define and submit their imagery requests.

"This information is integrated into daily communications with the station astronauts about their photo targets, so that crew members know what kind of photos would help the scientists, and when those areas will be visible through the windows of the station," said Cindy Evans, a NASA scientist who has been developing the online scientific resources.

Sweating Over the Next-Generation Life Support System

Marshall Space Flight Center employees are back at it—donating time and energy—exercising on treadmills, rowers, and bikes to test aspects of a life support system that could someday provide drinking water to people living on the Moon or Mars.

On Earth, nature provides the air we breathe, the water we drink, and natural resources that support life. In space, life support resources must be brought up with a crew, recycled or produced from resources available, or a combination of the two. For almost 20 years, NASA engineers at Marshall have led the design and development of the ISS life support system, called the Regenerative Environmental Control and Life Support System, or ECLSS.

In 2007, the ECLSS Oxygen Generation System was installed in the station's Destiny laboratory, and now, along with the Russian Elektron system, it will provide breathing air for future space station crews. The other component to ECLSS, the Water Recovery and Management System, takes in crew perspiration, respiration, and urine, and turns them into drinking water. The station's Water Recovery and Management System is planned to be delivered and installed in 2008.

Looking ahead to extended Moon missions, when re-supply will be over 240,000 miles away, Marshall engineers have assembled key aspects of the station's ECLSS waste water processor technology to explore how this system might work on a future lunar habitat. This redesigned hardware, the Exploration Water Recovery System, is a novel combination of proven air and water purification technologies and optimizes the treatment of various wastewater streams.

"To support human life on the Moon, we'll need robust and efficient life support systems that can work well without a large amount of consumables," said Monsi Roman, Exploration Life Support project manager. "Our hope is to mature current life support technologies to be able to minimize the amount of materials we need to bring up to space to support future crews."

For several weeks, Marshall tested this new hardware. The goal of this test was to examine the efficiency of the water processor to remove different types of contaminants from the wastewater. NASA engineers want to determine

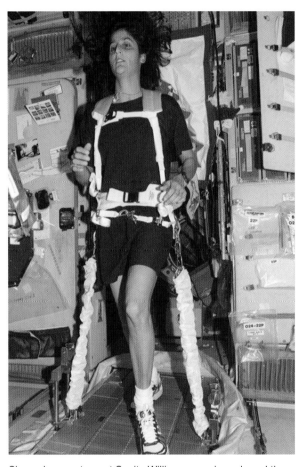

Shown here, astronaut Sunita Williams exercises aboard the International Space Station. The Exploration Water Recovery System collects perspiration, respiration, and urine and processes them into drinking water.

how to increase the system efficiency and extend the life of expendables needed to keep clean water flowing.

More than 50 employees participated in the Exploration Water Recovery System test. For the study, 20 employees exercised for an hour a day, generating water vapor through perspiration and respiration in the Regenerative ECLSS Module Simulator—a mockup of a space module filled with treadmills, a bicycle, rowing machine, and other exercise equipment. The men also donated urine as part of this test.

Before stepping into the module for a session, participants were provided with a T-shirt to wear, a towel for drying off, and a bottle of water or a sports energy drink to consume as they exercised. They weighed-in on a computerized scale, with the bottle of water in-hand. Sopping wet T-shirts and used towels are left hanging inside overnight to evaporate more sweat out of them. Participants brushed their teeth, wiped themselves down with wet towels and the men even shaved—simulating the daily routine of a station crew member—to get every bit of moisture into the atmosphere. Participants even microwaved meals inside the module to generate water vapor.

"We know this equipment can create water cleaner than water from municipal water systems here on Earth," said Keith Parrish, ECLSS Test Facility manager. "We hope we can refine the process so future crews will need fewer supplies to generate water for longer space missions—whether on the Moon or Mars."

Science Mission Directorate

The Science Mission Directorate engages the Nation's science community, sponsors scientific research, and develops and deploys satellites and probes in collaboration with NASA's partners around the world to answer fundamental questions requiring the view from and into space. The directorate seeks to understand the origins, evolution, and destiny of the universe and to understand the nature of the strange phenomena that shape it.

NASA Spacecraft Streams Back Surprises from Mercury

After a journey of more than 2 billion miles and 3½ years, NASA's MErcury Surface, Space ENvironment, GEochemistry and Ranging (MESSENGER) spacecraft made its flyby on Jan. 14, 2008, the first sent to orbit

MESSENGER's flyby of Mercury showed some of the major never-before-seen terrain in the inner solar system.

the planet closest to our Sun. The spacecraft's cameras and other sophisticated, high-technology instruments collected more than 1,200 images and made other science observations. Data included the first up-close measurements of Mercury since the Mariner 10 spacecraft's third and final flyby on March 16, 1975.

MESSENGER's flyby of Mercury gave scientists an entirely new look at a planet once thought to have characteristics similar to those of Earth's Moon. Researchers are amazed by the wealth of images and data that show a unique world with a diversity of geological processes and

a very different magnetosphere from the one discovered and sampled more than 30 years ago.

The spacecraft showed that Mercury has huge cliffs with structures snaking up hundreds of miles across the planet's face. These cliffs preserve a record of patterns of fault activity from early in the planet's history. The spacecraft also revealed impact craters that appear very different from lunar craters.

Instruments provided a topographic profile of craters and other geological features on the night side of Mercury. The spacecraft also discovered a unique feature that scientists dubbed "The Spider." This formation never has been seen on Mercury before and nothing like it has been observed on the Moon. It lies in the middle of a large impact crater called the Caloris basin and consists of more than 100 narrow, flat-floored troughs radiating from a complex central region.

Now that the spacecraft has shown scientists the full extent of the Caloris basin, its diameter has been revised upward from the Mariner 10 estimate of 800 miles to perhaps as large as 960 miles from rim to rim. The plains inside the Caloris basin are distinctive and more reflective than the exterior plains. Impact basins on the Moon have opposite characteristics.

The magnetosphere and magnetic field of Mercury during the flyby appeared to be different from the Mariner 10 observations. The spacecraft found the planet's magnetic field was generally quiet but showed several signatures indicating significant pressure within the magnetosphere.

Magnetic fields like Earth's and their resulting magnetospheres are generated by electrical dynamos in the form of a liquid metallic outer core deep in the planet's center. Of the four terrestrial planets, only Mercury and Earth exhibit such a phenomenon. The magnetic field deflects the solar wind from the Sun, producing a protective bubble around Earth that shields the surface of our planet from those energetic particles and other sources farther out in the galaxy.

Similar variations are expected for Mercury's magnetic field, but the precise nature of its field and the time scales for internal changes are unknown. The next two flybys and the year-long orbital phase will shed more light on these processes.

The spacecraft's suite of instruments has provided insight into the mineral makeup of the surface terrain and detected ultraviolet emissions from sodium, calcium, and hydrogen in Mercury's exosphere. It also has explored the sodium-rich exospheric "tail," which extends more than 25,000 miles from the planet.

Astronomers Detect First Organic Molecule on an Exoplanet

A team of astronomers led by Mark Swain of NASA's Jet Propulsion Laboratory (JPL) has made the first detection ever of an organic molecule in the atmosphere of a Jupiter-sized planet orbiting another star. The breakthrough, made with NASA's Hubble Space Telescope, is an important step in eventually identifying signs of life on a planet outside our solar system.

The molecule found by Hubble is methane, which can play a key role in prebiotic chemistry—the chemical reactions considered necessary to form life as we know it. This discovery proves that Hubble and upcoming space missions, such as NASA's James Webb Space Telescope, can detect organic molecules on planets around other stars by using spectroscopy, which splits light into its components to reveal the "fingerprints" of various chemicals.

"This is a crucial stepping stone to eventually characterizing prebiotic molecules on planets where life could

Extrasolar planet HD 189733b with its parent star peeking above its top edge. Astronomers used the Hubble Space Telescope to detect methane and water vapor in the Jupiter-sized planet's atmosphere.

exist," said Swain, lead author of a paper that appeared in the March 20, 2008 issue of Nature.

The discovery comes after extensive observations made in May 2007 with Hubble's Near Infrared Camera and Multi-Object Spectrometer. It also confirms the existence of water molecules in the planet's atmosphere, a discovery made originally by NASA's Spitzer Space Telescope in 2007. "With this observation there is no question whether there is water or not—water is present," said Swain.

The planet, called HD 189733b, is located 63 light-years away in the constellation Vulpecula. Though it is now known to have methane and water, the planet is so massive and so hot it is considered an unlikely host for life; HD 189733b is so close to its parent star it takes just over 2 days to complete an orbit, and its atmosphere

swelters at 1,700 °F, about the same temperature as the melting point of silver.

Though the star-hugging planet is too hot for life as we know it, "This observation is proof that spectroscopy can eventually be done on a cooler and potentially habitable Earth-sized planet orbiting a dimmer red dwarf-type star," Swain said. The ultimate goal of studies like these is to identify prebiotic molecules in the atmospheres of planets in the "habitable zones" around other stars, where temperatures are right for water to remain liquid rather than freeze or evaporate away.

"These measurements are an important step to our ultimate goal of determining the conditions, such as temperature, pressure, winds, clouds, etc., and the chemistry on planets where life could exist. Infrared spectroscopy is

really the key to these studies because it is best matched to detecting molecules," said Swain.

Cassini Flies through Watery Plumes of Saturn Moon

NASA's Cassini spacecraft performed a flyby of Saturn's moon Enceladus on March 12, 2008, flying about 15 kilometers per second (32,000 mph) through icy water geyser-like jets. The spacecraft snatched up precious samples that might point to a water ocean or organics inside the little moon.

Scientists believe the geysers could provide evidence that liquid water is trapped under the icy crust of Enceladus. The geysers emanate from fractures running along the moon's South Pole, spewing out water vapor at approximately 400 meters per second (800 mph).

The new data provide a much more detailed look at the fractures that modify the surface and will give a significantly improved comparison between the geologic history of the moon's North and South Poles.

New images show that compared to much of the Southern Hemisphere on Enceladus—the south polar region in particular—the north polar region is much older and pitted with craters of various sizes. These craters are captured at different stages of disruption and alteration by tectonic activity, and probably from past heating from below. Many of the craters seem sliced by small parallel cracks that appear to be ubiquitous throughout the old cratered terrains on Enceladus.

These new images are showing us in great detail how the moon's North Pole differs from the South, an important comparison for working out the moon's obviously complex geological history.

The flyby was designed so that Cassini's particle analyzers could dissect the "body" of the plume for information on the density, size, composition, and speed of the particles. Among other things, scientists will use the data gathered to figure out whether the gasses from the plume match the gasses that make up the halo of particles

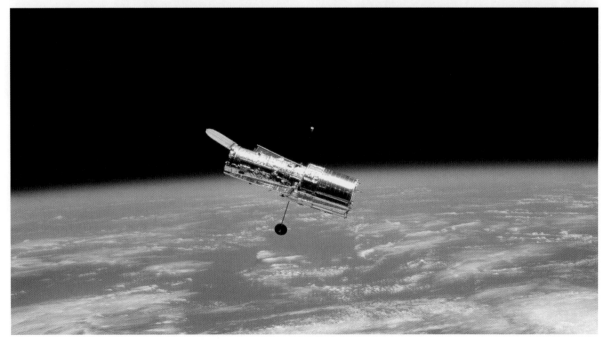

NASA's Hubble Space Telescope has helped astronomers identify organic molecules in the atmosphere of a planet orbiting another star.

around Enceladus. This may help determine how the plumes formed.

During Cassini's closest approach, two instruments were collecting data—the cosmic dust analyzer and the ion and neutral mass spectrometer. An unexplained software hiccup with Cassini's cosmic dust analyzer instrument prevented it from collecting any data during the closest approach, although the instrument did get data before and after the approach. During the flyby, the instrument was switching between two versions of software programs. The new version was designed to increase the ability to count particle hits by several hundred hits per second. The other four fields and particles instruments, in addition to the ion and neutral mass spectrometer, did capture all of their data, which will complement the overall composition

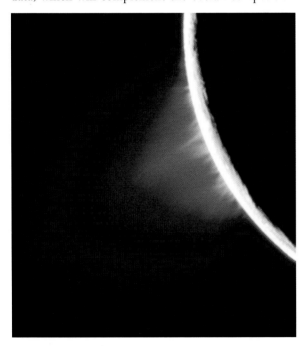

NASA's Cassini spacecraft performed a flyby of Saturn's moon Enceladus, snatching up precious samples in the icy geyser-like jets.

Two elements comprise the spacecraft: the Cassini orbiter and the Huygens probe. In 2004, Cassini-Huygens reached Saturn and its moons.

studies and elucidate the unique plume environment of Enceladus.

Cassini's instruments discovered evidence for the geyser-like jets on Enceladus in 2005, finding that the continuous eruptions of ice water create a gigantic halo of ice dust and gas around Enceladus, which helps supply material to Saturn's E-ring.

This was the first of four Cassini flybys of Enceladus this year, during which, the spacecraft came within 50 kilometers (30 miles) of the surface at closest approach, 200 kilometers (120 miles) while flying through the plume. Future trips may bring Cassini even closer to the surface of Enceladus. Cassini completes its prime mission, a 4-year tour of Saturn, this summer. From then on, a proposed extended mission would include seven more Enceladus flybys, beginning in late summer.

NASA Spacecraft Photographs Avalanches on Mars

A NASA spacecraft in orbit around Mars took the first-ever image of active avalanches near the Red Planet's North Pole. The image showed tan clouds billowing away from the foot of a towering slope, where ice and dust have just cascaded down.

The High Resolution Imaging Science Experiment (HiRISE) on NASA's Mars Reconnaissance Orbiter took the photograph on February 19, 2008, and this is one of approximately 2,400 HiRISE images recently released. The full image reveals features as small as a desk in a strip of terrain 3.7 miles wide and more than 10 times that long, at 84 degrees north latitude. Reddish layers known to be rich in water ice make up the face of a steep slope more than 2,300 feet tall, running the length of the image.

More ice than dust probably makes up the material that fell from the upper portion of the scarp. Imaging of the site during coming months will track any changes in the new deposit at the base of the slope. That will help researchers estimate what proportion is ice.

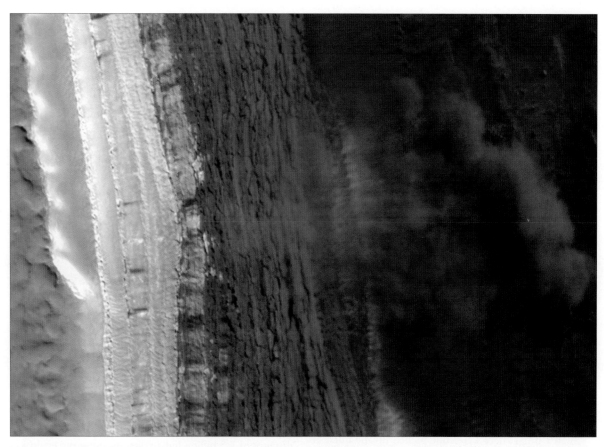

In February 2008, the High Resolution Imaging Science Experiment (HiRISE) on NASA's Mars Reconaissance Orbiter captured this image of dust billowing away from an avalanche.

Another notable HiRISE image released showed a blue crescent Earth and its moon, as seen by the Mars Reconnaissance Orbiter. The west coast of South America is visible in the photo. Still other images allow viewers to explore a wide variety of Martian terrains, such as dramatic canyons and rhythmic patterns of sand dunes.

The camera is one of six science instruments on the orbiter. The spacecraft reached Mars in March 2006 and has returned more data than all other current and past missions to Mars combined.

NASA Satellite Detects Record Gamma Ray Burst Explosion Halfway Across Universe

A powerful stellar explosion detected March 19, 2008, by NASA's Swift satellite shattered the record for the most distant object that could be seen with the naked eye.

The explosion was a gamma ray burst. Most gamma ray bursts occur when massive stars run out of nuclear fuel. Their cores collapse to form black holes or neutron

Swift's Burst Alert Telescope picked up this gamma ray burst, one of the most luminous explosions in the universe.

stars, releasing an intense burst of high-energy gamma rays and ejecting particle jets that rip through space at nearly the speed of light like turbocharged cosmic blowtorches. When the jets plow into surrounding interstellar clouds, they heat the gas, often generating bright afterglows. Gamma ray bursts are the most luminous explosions in the universe since the Big Bang.

Swift's Burst Alert Telescope picked up the burst at 2:12 a.m. Eastern Time, March 19, and then pinpointed the coordinates. Telescopes in space and on the ground quickly moved to observe the afterglow. The burst is named "GRB 080319B," because it was the second gamma ray burst detected that day.

Swift's other two instruments, the X-ray Telescope and the Ultraviolet/Optical Telescope, also observed brilliant afterglows. Several ground-based telescopes saw the afterglow brighten to visual magnitudes between 5 and 6 in the logarithmic magnitude scale used by astronomers. The brighter an object is, the lower its magnitude number. From a dark location in the countryside, people with

normal vision can see stars slightly fainter than magnitude 6. That means the afterglow would have been dim, but visible to the naked eye.

Later that evening, the Very Large Telescope, in Chile, and the Hobby-Eberly Telescope, in Texas, measured the burst's redshift at 0.94. A redshift is a measure of the distance to an object. A redshift of 0.94 translates into a distance of 7.5 billion light-years, meaning the explosion visible now took place 7.5 billion years ago, a time when the universe was less than half its current age and Earth had yet to form. This is more than halfway across the visible universe.

GRB 080319B's optical afterglow was 2.5 million times more luminous than the most luminous supernova ever recorded, making it the most intrinsically bright object ever observed by humans in the universe. The most distant previous object that could have been seen by the naked eye is the nearby galaxy M33, a relatively short 2.9 million light-years from Earth.

Analysis of GRB 080319B is underway, because astronomers don't know why this burst and its afterglow were so bright. One possibility is the burst was more energetic than others, perhaps because of the mass, spin, or magnetic field of the progenitor star or its jet. Or perhaps it concentrated its energy in a narrow jet that was aimed directly at Earth.

Swift is managed by Goddard Space Flight Center. It was built and is being operated in collaboration with Penn State, the Los Alamos National Laboratory, and General Dynamics, in the United States; the University of Leicester and Mullard Space Science Laboratory, in the United Kingdom; Brera Observatory and the Italian Space Agency, in Italy; plus partners in Germany and Japan.

First-Ever 3-D Images of the Sun

NASA's twin Solar Terrestrial Relations Observatory (STEREO) spacecraft made the first three-dimensional images of the Sun. The new view will greatly aid scientists' ability to understand solar physics and thereby improve space weather forecasting.

The improvement with STEREO's 3-D view is like going from a regular X-ray to a 3-D CAT scan in the medical field.

The STEREO spacecraft were launched October 25, 2006. On January 21, 2007, they completed a series of complex maneuvers, including flying by the Moon, to position the spacecraft in their mission orbits. The two observatories are now orbiting the Sun, one slightly ahead of Earth and one slightly behind, separating from each other by approximately 45 degrees per year. Just as the slight offset between a person's eyes provides depth perception, the separation of these spacecraft allow 3-D images of the Sun.

Violent solar weather originates in the Sun's atmosphere, or corona, and can disrupt satellites, radio communication, and power grids on Earth. The corona resembles wispy smoke plumes, which flow outward along the Sun's tangled magnetic fields. It is difficult for

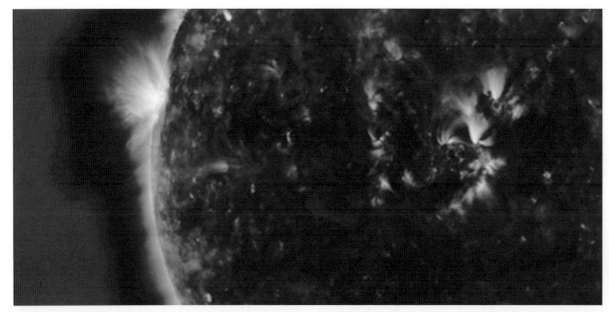

NASA's twin STEREO spacecraft captured the first three-dimensional images of the Sun, which are aiding scientists' ability to understand solar physics.

science center. The Johns Hopkins University Applied Physics Laboratory designed and built the spacecraft and is responsible for mission operations. The STEREO imaging and particle-detecting instruments were designed and built by scientific institutions in the United States, United Kingdom, France, Germany, Belgium, Netherlands, and Switzerland.

Phoenix Mars Lander Analyzes Particles for Water

Launched in August 2007, the Phoenix Mars Lander Mission is the first in NASA's Mars Scout class, combining legacy and innovation in a framework of a true partnership: government, academia, and industry. Phoenix is designed to study the history of water and habitability potential in the Martian arctic's ice-rich soil.

scientists to tell which structures are in front and which are behind.

With STEREO's 3-D imagery, scientists will be able to discern where matter and energy flows in the solar atmosphere much more precisely than with the 2-D views previously available. This will help scientists understand the complex physics going on.

STEREO's depth perception also will help improve space weather forecasts. Of particular concern is a destructive type of solar eruption called a Coronal Mass Ejection (CME). CMEs are eruptions of electrically charged gas, called plasma, from the Sun's atmosphere. A CME cloud can contain billions of tons of plasma and move at a million miles per hour.

The CME cloud is laced with magnetic fields, and CMEs directed toward Earth smash into our planet's magnetic field. If the CME magnetic fields have the proper orientation, they dump energy and particles into

Earth's magnetic field, causing magnetic storms that can overload power line equipment and radiation storms that disrupt satellites.

Satellite and utility operators can take precautions to minimize CME damage, but they need an accurate forecast of when the CME will arrive. To do this, forecasters need to know the location of the front of the CME cloud. STEREO will allow scientists to accurately locate the CME cloud front. Knowing where the front of the CME cloud is will improve estimates of the arrival time from within a day or so to just a few hours. STEREO also will help forecasters estimate how severe the resulting magnetic storm will be.

STEREO's first 3-D images were provided by JPL. STEREO is the third mission in NASA's Solar Terrestrial Probes program within NASA's Science Mission Directorate. The Goddard Science and Exploration Directorate manages the mission, instruments, and

The Surface Stereo Imager captured this image on Sol 11 (June 5, 2008). The robotic arm carries a soil sample towards the partially open door of the Thermal and Evolved-Gas Analyzer's number four cell, or oven.

In this artist rendition, the Phoenix lander begins to dig a trench through the upper soil layer on the arctic plains of Mars. The polar water ice cap is shown in the far distance. Corby Waste of the Jet Propulsion Laboratory created this image.

The Phoenix Mars Lander's Surface Stereo Imager took this image of the solar panel and the lander's Robotic Arm while facing west during Phoenix's Sol 16, or June 10, 2008. The robotic arm is delivering the sample in the scoop to the Optical Microscope.

Led by principal investigator Peter Smith, of the University of Arizona's Lunar and Planetary Laboratory, the science team aims to answer the following questions: 1) Can the Martian arctic support life? 2) What is the history of water at the landing site? 3) How is the Martian climate affected by polar dynamics?

To answer these questions, Phoenix uses some of the most sophisticated and advanced technology ever sent to Mars. A robust robotic arm built by JPL digs through and delivers soil and ice samples to the mission's experiments.

On the deck, miniature ovens and a mass spectrometer provide chemical analysis of trace matter, and a chemistry lab characterizes the soil and ice chemistry. Imaging systems designed by the University of Arizona, University of Neuchâtel (Switzerland), Max Planck Institute (Germany), and Malin Space Science Systems (California) will render unprecedented, detailed views of Mars. The lander's meteorological station furnished by the Canadian Space Agency marks the first significant involvement of Canada in a mission to Mars.

From the Science Operations Center in Tucson, the Phoenix science and engineering teams began commanding the lander once it arrived safely on Mars and began transmitting data to Earth. This powerful team has high hopes for this to be the first mission to "touch" and examine water on Mars—ultimately, to pave the way for future robotic missions and possibly, human exploration.

On June 11, 2008, two instruments on the lander deck—a microscope and a Thermal and Evolved-Gas Analyzer (TEGA), or "oven" instrument—began inspecting

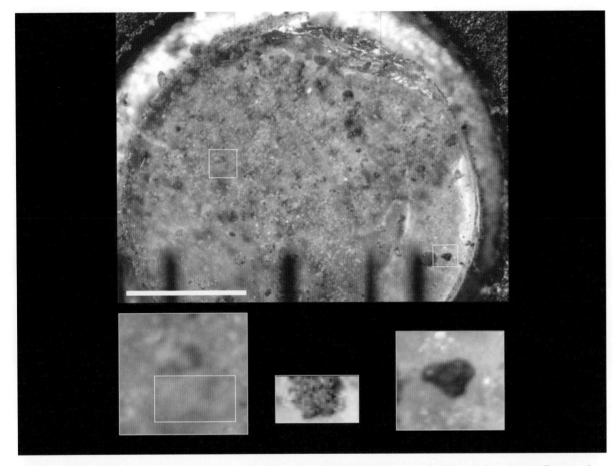

Three close-ups of Martian soil appear in this composite from the Optical Microscope on the Phoenix Mars lander. The top of the composite particle (left box) has a green tinge, possibly indicating olivine. The bottom of the particle has been reimaged in black and white (middle box), showing that this is a clump of finer particles. The right box shows a rounded, glassy particle. The scale bar is 1 millimeter. This was taken on June 11, 2008 or Phoenix Sol 17.

soil samples delivered by the scoop on Phoenix's robotic arm. The sample includes some larger, black, glassy particles as well as smaller reddish ones. The fine particles in the soil sample closely resemble particles of airborne dust examined earlier by the microscope.

"The oven is working very well and living up to our expectations," said Phoenix co-investigator and TEGA team leader, Bill Boynton, of the University of Arizona. Phoenix has eight separate tiny ovens to bake the samples at three different temperature ranges and sniff the soil to look for volatile ingredients (such as water). Studying dust on Mars helps scientists understand atmospheric dust on Earth, which is important because dust is a significant factor in global climate change.

NASA's Aeronautics Research Mission Directorate, led by Dr. Jaiwon Shin, conducts cutting-edge, fundamental research in traditional and emerging disciplines to help transform the Nation's air transportation system, and to support future air and space vehicles. Its goals are to improve airspace capacity and mobility, improve aviation safety, and improve aircraft performance while reducing noise, emissions, and fuel burn.

NASA Researchers Design New Chip

NASA researchers have designed and built a new circuit chip that can take the heat like never before. In the past, integrated circuit chips could not withstand more than a few hours of high temperatures before failing. This silicon carbide (SiC) chip exceeded 5,000 hours of continuous operation at 500 °C—a breakthrough that represents a 500-fold increase in what had previously been achieved. Such highly durable integrated packaging is being developed to enable extremely functional but physically small circuitry for hot sections of jet engines.

Electronics to control combustion affect people everyday. In 100-200 °C high-temperature environments, electronic chips made from silicon are crucial to modern jet and automobile engine performance. Look under the cowling of a modern jet engine or under the hood of a modern automobile, and you will find wires leading to and from electronic sensors and controls to electronic engine control unit circuitry. But silicon cannot function durably in the much hotter 400-500 °C environments where even more capable electronic sensing and control is needed.

These new circuit chips will let us get closer to the combustion source and enable better sensing and control with fewer wires, connectors, and cooling requirements for the combustion sensing/control subsystems. The first use for this capability will likely be in the military and aerospace markets since they are much less cost-sensitive than other markets.

In the future, such electronics will enhance sensing and control of the combustion process that could lead to improved safety and fuel efficiency as well as reduced emissions from jet engines. Similar benefits are also possible for automotive engines, oil and natural gas well drilling, and anything requiring long-lasting electronic circuits in very hot environments, including robotic exploration on the hostile surface environment of Venus.

The next step in SiC chip development is to greatly increase the single-chip transistor count from less than 10 transistors upward toward 100 to 1,000 transistors per chip, which in turn would enable much greater 500 °C integrated circuit functionality beyond simple signal amplification. Since much of the knowledge on

This breakthrough silicon carbide chip exceeded 5,000 hours of continuous operation at 500 °C.

how to put more transistors on a single chip already exists from silicon-integrated circuit technology, we expect the scale-up process for durable 500 °C SiC-integrated circuitry to occur within 2 to 4 years.

NASA Increases Aviation Safety

Advancements in information technologies, such as data mining, are enabling the aviation community to pursue a more proactive approach for preventing accidents. In FY2007, NASA, in collaboration with the Federal Aviation Administration (FAA) and the commercial aviation community, established the Aviation Safety Information Analysis and Sharing (ASIAS) system, which is being used to integrate and analyze large sources of operational data in order to detect anomalies and/or dangerous trends before an accident occurs. NASA has also developed three new data-mining methods that are intended to help the aviation community to more efficiently collect and analyze the ASIAS data:

- The System-Level Morning Report (SLMR) helps airlines to automatically uncover small clusters of operationally atypical flights using data from operator activities such as airline Flight Operational Quality Assurance (FOQA) programs. These small clusters may be indicative of underlying safety hazards. The SLMR does not require any pre-specification of what constitutes "typical" flying patterns. This contributes to better efficiency and the ability to uncover unexpected safety issues. The SLMR could potentially be incorporated into FOQA as a routine processing service that would enable each airline to compare itself to the system-level results.

- SequenceMiner identifies anomalous sequences of switch activations or, more precisely, any unusual discrete parameter patterns within a flight phase. Detected anomalies include more or less frequent activations than expected and unusual switch sequencing. SequenceMiner can potentially identify

problems associated with mode confusion, equipment troubleshooting, abnormal situation response, etc.

- Mariana auto-classifies aviation safety reports based on the contents of report narratives and a limited number of fixed fields such as the event flight phase. The classifications could potentially relate to event types, contributing factors, and other features of interest. Mariana could potentially produce more consistent Aviation Safety Action Program report classifications for airlines, promoting effective data integration across multiple users.

NASA's Airspace Systems Program, the FAA's Air Traffic Organization (ATO), and the Joint Planning and Development Office (JPDO) are working collaboratively to establish a process to transfer technologies from fundamental research and development (R&D) into implementation for the next-generation air transportation system, or NextGen. This process has top-level commitment from Shin and Victoria Cox, ATO's vice president for Operations Planning Services. A coordinating committee that includes both FAA and NASA representatives oversees four research transition teams that are organized around the NextGen Concept of Operations framework. This framework connects the FAA's Operational Evolution Partnership elements to NASA's aeronautics research portfolio. The JPDO plays an important role in this transfer in that they keep everyone informed on the progress of the Integrated Work Plan. The teams are working to plan near-term R&D transition areas such as surface management, and in long-term transition areas such as dynamic airspace allocation.

Increasing Turbine Engine Efficiency

In the past, improvements in turbine engine efficiency, performance, noise, and emissions, have been achieved through combinations of new component designs, as well as new materials with higher temperature capability and lower weight. In the future, however, one of the most

Newly developed high-temperature shape memory alloys are helping to reduce noise and improve efficiency in jet engines. NASA has sped the transfer of this technology from the laboratory to practical use worldwide.

exhibiting temperature capability between 200-400 °C, while still maintaining high work capability levels.

One example of the successful use of high-temperature SMAs is known as an adaptive chevron. Chevrons placed on either the fan or core exhaust of jet engines have been shown to be very effective in reducing noise during the takeoff of commercial aircraft. However, conventional fixed-geometry chevrons represent a tradeoff in design that reduces noise but also imposes a small, but significant, performance penalty on aircraft engines. The solution would be an adaptive-geometry chevron that could mitigate noise during takeoff and be retracted during cruise so as not to affect performance. However, there are several daunting challenges to developing a successful adaptive-geometry chevron, especially for core exhaust applications. These include the need for very high actuation forces, a very limited volume for actuator placement, and high operating temperatures. Initial testing indicates that these challenges may be overcome through the use of a new chevron design developed by Continuum Dynamics Inc., that uses a high-temperature SMA alloy developed and supplied by NASA's Glenn Research Center. This resulted in an overall concept with deflection performance that is at least five times more efficient than current approaches.

Another example of the successful use of high-temperature SMAs is known as active flow control. A compressor's aerodynamics can improve efficiency but may be limited by the onset of compressor stall conditions. A strategy of advanced sensors that can detect the onset of incipient stall, which would then trigger small changes to the flow geometry, allows for a compressor that has both improved efficiency plus tolerance against stall conditions. A design utilizing a high-temperature SMA wire to actuate a control rod which is inserted into the air flow has been designed and successfully demonstrated. Two subsequent design iterations have reduced the size of the SMA actuator by a factor of 10, thus making the concept even more attractive. ❖

effective ways to increase engine performance will be to replace various static structures with adaptive or reconfigurable components.

Designs for everything from adaptive inlets, nozzles, flaps, and other control surfaces, to variable-geometry chevrons, reconfigurable blades, and active hinges for the operation of various doors and panels have been developed. Many of these concepts have been envisioned in the past, but could not be achieved because they relied upon electric, hydraulic, or pneumatic actuators that brought excessive weight penalties. NASA's approach currently is

to utilize advanced shape memory alloys (SMAs) in order to achieve these breakthroughs.

The most viable SMAs have been based on binary nickel titanium (NiTi), but this class of SMA has a very low-temperature capability, generally in the range of -100 to 80 °C. Many of the envisioned applications in aeronautics require SMAs that have temperature capability far in excess of this level but that can still generate significant work output. The results of a 5-year development effort are now reaching maturity. Newly developed alloys, formed by ternary and quaternary additions to NiTi, are

Education News

Using the inspirational pursuit of space exploration to spark the imagination of our youth is critical to keeping this Nation competitive and creating a scientifically literate populace. Toward that end, NASA has a tradition of investing in programs and activities that engage students, educators, families, and communities in exploration and discovery. Read about the innovative education activities NASA is creating in support of its missions and future workforce needs.

Education News

Building the Foundation for the Next Era of Exploration

This year, NASA celebrates 50 years of exploring exciting frontiers that have led to new horizons of opportunity. As NASA implements the U.S. Space Exploration Policy, carrying humans back to the Moon, on to Mars, and beyond, the Agency is also working to lay the educational groundwork that will make this ongoing journey possible.

NASA is embarking on a program of exploration that will continue for decades, requiring the dedication and ingenuity of scientists and engineers today and for generations to come. To ensure those future explorers will be ready to continue the journey, NASA is working with one of its most important partners—educators.

The Agency recognizes the importance of educators' contributions in making our work possible, and is dedicated to supporting them in the disciplines of science, technology, engineering, and mathematics (STEM). NASA provides formal and informal educators unique resources and development opportunities to strengthen the overall teaching of STEM subjects. In the summer of 2007, Mission Specialist and Educator Astronaut Barbara R. Morgan captured students' imaginations as she flew to space aboard the Space Shuttle Endeavour on an assembly mission to the International Space Station (ISS). In the winter of 2008, Mission Specialists and Educator Astronauts Ricky Arnold and Joe Acaba are journeying to the ISS to perform spacewalks during mission STS-119. Several exciting education activities are planned around the mission.

NASA also aims to attract and retain students in STEM disciplines through a progression of educational opportunities for students, teachers, and faculty, and to build strategic partnerships and links between formal and informal education providers that promote STEM literacy and awareness of NASA's mission. The Office of Education promotes education as an integral component of every major NASA research and development mission. NASA, with industry and university engineers and scientists, is sharing knowledge and experience with students and educators as they study Earth and the universe using the latest aerospace research methods.

These efforts are accomplished through collaboration among NASA's Office of Education, Mission Directorates, and Field Centers, as well as other Federal agencies

NASA aims to inspire and engage the next generation of scientists and engineers who will continue the Agency's missions in decades to come.

engaged in education activities, and various public and private partners. NASA's Office of Education is committed to providing opportunities for all students to explore and experience unique space and aeronautics content, to interact with innovative engineers and scientists, and to see state-of-the-art facilities. The Agency remains committed to engaging underrepresented and underserved communities of students, educators, and researchers.

As NASA Administrator Michael Griffin explained, "The greatest contribution that NASA makes in educating the next generation of Americans is providing worthy endeavors by which students will be inspired to study difficult subjects like math, science, and engineering, because they too share the dream of exploring the cosmos."

Engaging Students in the Digital Age

NASA aims to continue contributing to the development of the Nation's STEM workforce of the future by identifying and developing the critical skills and capabilities needed to support the U.S. Space Exploration Policy. Reaching today's students calls for meeting them where they congregate, and often that is through digital and electronic environments.

In 2007, a team at Marshall Space Flight Center began the process of researching new ways to engage students using technology. The team reviewed research regarding podcasts, which are media files that are distributed over the Internet for playback on personal computers or portable devices, such as an MP3 player or cell phone. Research showed that the demographic most likely to download a podcast is between 18 and 24 years of age (Nielsen Analytics, April 2006), that 29 percent of Americans who have listened to or viewed a podcast are between 12 and 24 years of age (Arbitron/Edison Media Research, March 2007), that nearly half of Gen X and Gen Y (ages 14-43) download podcasts at least once a week (Forrester, February 2007), and that roughly half of teenagers own an iPod or other portable digital music player (Arbitron, July 2006). The team reviewed multiple styles of podcast,

NASA Student Opportunities podcasts feature interviews with students who have had the experience of working with NASA through the Agency's many learning activities.

Middle school students from across the country competed in a NASA-sponsored event on designing futuristic city concepts.

ultimately settling on an interview-style format where a student is interviewed about their experiences supporting NASA missions.

The resulting NASA Student Opportunities podcast series debuted in February 2007 and provides timely information to students on experiences of their peers, while also promoting future educational opportunities. The weekly podcasts are radio-style shows that feature interviews with students who have participated in NASA learning opportunities. These shows allow potential participants to hear first-hand what it is like to be a student member of a NASA team, and the free podcast also provides up-to-date information on approaching application deadlines. Forty-two episodes of NASA Student Opportunities were created between February and December 2007, and as of

March 2008, about 650,000 downloads have been collectively logged.

Teams Face Off at NASA in Future City Competition

The bright and fertile minds of middle school students across the United States have cultivated their visions of what future cities must look like in order to support humankind's growing infrastructure needs. During February 2008, they put their conceptualizations to the test in a competition focused on nanotechnology.

Over 30,000 students from 1,000 schools in 40 regions participated in the 16th annual National Engineers Week Future City Competition, dreaming up the most practical application of built-in nanotechnologies to monitor

parts of a city's infrastructure. Small, tightly knit teams of students, along with their teacher and engineer mentors, first created their future city digitally, using SimCity 3000 software. They then transformed their ideas into reality by sculpting a large table-top model using recycled materials that cost no more than $100. Each team was judged for their models, an essay, and a presentation that defended their approach to resolving monitoring issues for tomorrow's cities using nanotechnology. The winning team was from Heritage Middle School, in Westerville, Ohio.

STS-118 Provides Space Harvest for Classrooms

NASA's education quest was taken to new heights in August 2007 as Mission Specialist and Educator Astronaut Barbara R. Morgan made her first flight into space. NASA flew two education payloads aboard Space Shuttle Endeavour on mission STS-118 as part of NASA and the International Technology Education Association's Engineering Design Challenge for students. As part of the

Mission Specialist and Educator Astronaut, Barbara R. Morgan, talks to students prior to STS-118.

Working under the supervision of their teachers, students designed, built, tested, redesigned, and rebuilt models that met specified design criteria. As they improved their designs, students employed the same analytical skills as engineers. The design challenge culminated in the classroom, with each student team preparing a poster that described the process and results of their work.

Some students were treated to a special day in November 2007, when NASA, the U.S. National Arboretum, the U.S. Department of Education, and The Herb Society of America hosted an out-of-this-world school field trip to the U.S. National Arboretum, in Washington, DC, as part of International Education Week.

Seventy-two students from the Arlington Science Focus School, in Arlington, Virginia, presented to a panel of experts the lunar plant growth chambers they created as part of the Engineering Design Challenge. The panel included engineers, plant specialists and STS-118 crewmembers Scott Kelly, commander; Barbara R. Morgan, educator astronaut and mission specialist; and Dave Williams, Canadian Space Agency mission specialist.

These Virginia students used their design chambers to grow cinnamon basil seeds that were flown during the STS-118 mission. The soil for the seeds came from a unique source—the Washington Nationals ballpark. Arcillite, a claylike substance used in ballpark field maintenance, provides optimal growing conditions in a growth chamber.

As of spring 2008, over 30,000 sets of Earth and space seeds were distributed to classrooms nationwide that registered for the Engineering Design Challenge. Attendees at professional teacher conferences in 2008 such as the National Science Teachers Association Conference, in Boston, were also given the opportunity to register for the Engineering Design Challenge and receive seeds from space.

challenge, students in grades K-12 planned, designed, and built their version of a lunar plant growth chamber. They validated their chambers using space-exposed seeds and Earth-based control seeds.

The first payload consisted of two small, collapsible plant growth chambers and the associated hardware to conduct a 20-day plant germination investigation. During the investigation, crewmembers maintained the plants and captured images of plant growth.

Approximately 10 million basil seeds made up the second payload. The seeds flew into space in SPACEHAB, a pressurized, mixed-cargo carrier that supports various quantities, sizes, and locations of experiment hardware; remained throughout the mission; and returned to Earth, having stayed on the shuttle during the entire mission. After the mission, the seeds were distributed to students and educators as part of the Engineering Design Challenge.

As of spring 2008, over 30,000 sets of Earth and space seeds were distributed to classrooms nationwide that registered for the Engineering Design Challenge.

International Space Station as Educational Resource

The ISS is the largest and most complex space vehicle ever built. Planned for completion in 2010, the space station will provide a home for laboratories equipped with a wide array of resources to develop and test the technologies needed for future generations of space exploration.

In 2006, NASA asked a range of Federal agencies with responsibilities in education to participate in the ISS Education Coordination Working Group, charged with developing a strategy for using the ISS as an educational asset. The initial report from the task force, delivered in December 2006, affirmed that there was serious interest on the part of Federal agencies in use of the ISS.

NASA's report, "An Opportunity to Educate: ISS National Laboratory," presents a plan to validate the task force's strategy for using ISS resources and accommodations as a venue to engage, inspire, and educate students, teachers, and faculty in the STEM areas. Information about current NASA and non-Agency programs aimed to increase STEM achievement is included in the report. For the demonstration phase of the plan, 11 organizations submitted varied candidate demonstration projects which cover STEM subjects and convey the possibilities inherent in the ISS National Laboratory concept.

In June 2008, the document was finalized and submitted to Congress.

'Pete Conrad Spirit of Innovation Award' Winners

On January 18, 2008, at NASA Headquarters, NASA Deputy Administrator Shana Dale recognized the first winners of the "Pete Conrad Spirit of Innovation Award" during the unveiling of a traveling exhibit titled "Spirit of Innovation."

To compete for the $10,000 award, high school students were challenged to create concepts that could accelerate the personal space flight industry through

Michael Hakimi and Talia Nour-Omid from Los Angeles won first prize in the "Pete Conrad Spirit of Innovation Award" competition at the X PRIZE Cup in October 2007 and were honored in January 2008. Their winning idea was a device to monitor a human's vital signs while in space. (From the left: Michael Hakimi; Dr. Bernice Alston, deputy assistant administrator for planning, policy, and evaluation, NASA Office of Education; Talia Nour-Omid; and Doug Comstock, director of NASA's Innovative Partnerships Program.)

graphical representations, technical documents, and business plans. Winners were chosen during the X PRIZE Cup, administered by the X PRIZE Foundation under an educational grant from NASA, in New Mexico, October 2007.

The first-place team, "Michael and Talia" from Los Angeles, helped unveil the traveling exhibit that features their winning entry. The exhibit also showcases the 2007 competition, which was held at the X PRIZE Cup at Holloman Air Force Base, in Alamogordo, New Mexico. Michael Hakimi and Talia Nour-Omid developed an idea for a device that would effectively monitor all of a human being's vital signs while in space. The winning team received a $5,000 grant for their school and a trophy presented by Nancy Conrad, wife of the late Apollo astronaut Pete Conrad and creator of the prize, and

Erik Lindbergh, X PRIZE Foundation Trustee, great-grandson of Charles Lindbergh and designer and sculptor of the First Prize trophy.

NASA's Experimental Program to Stimulate Competitive Research

NASA awarded more than $17 million to institutions nationwide to help make significant contributions to the research and technology priorities of the Agency. The selections were part of NASA's Experimental Program to Stimulate Competitive Research (EPSCoR). The EPSCoR is designed to assist states in establishing an academic research enterprise directed towards a long-term, self-sustaining, and competitive capability that will contribute to the states' economic viability and development. EPSCoR enables continuing education, training, and workshops important to NASA's mission. The program assists in developing partnerships between NASA research assets, academic institutions, and industry.

A total of 23 proposals were selected for funding, representing: Alabama, Alaska, Arkansas, Idaho, Kentucky, Louisiana, Maine, Montana, Nebraska, New Hampshire, New Mexico, Nevada, Oklahoma, Puerto Rico, South Carolina, South Dakota, Vermont, and West Virginia. Winning proposals were selected through a merit-based, peer-reviewed competition. Proposals span the spectrum of NASA research priorities, from developing new wheels for rovers on extra-terrestrial exploration, to new batteries and fuel cells, to developing new bonding materials that will enhance aircraft safety.

Montana Launches into the Future

Explorer-1 was launched 50 years ago, on January 31, 1958, after the Soviets sent up Sputnik on October 4, 1957. The U.S. satellite observed the Van Allen radiation belt around Earth, a discovery said to be the first major scientific breakthrough of the space age. Montana State University (MSU) students who only know about the

Each year, students compete in the lunar rover challenges, hosted by the U.S. Space and Rocket Center, in Huntsville, Alabama.

Students Navigating the Moonscape

Each year around April, a half-mile of paths at the U.S. Space and Rocket Center, in Huntsville, Alabama, transform into a harsh lunar landscape. The course tested the engineering savvy and physical endurance of about 400 high school and college students on 68 teams who converged in April 2008 for NASA's 15th annual Great Moonbuggy Race.

The students, hailing from 20 states, Puerto Rico, Canada, India, and Germany, raced lightweight moonbuggies they designed, based on the original lunar rovers first used during the Apollo 15 Moon mission in 1971. They faced 17 unique course obstacles, built of plywood and old tires, and covered with 20 tons of gravel and 5 tons of sand that had been shaped into Moon-like ridges, craters, sandy basins, and lava-etched "rilles."

The race challenged students to design a vehicle that addresses a series of engineering problems similar to those faced by the team that designed the original Apollo-era lunar rover. The basic challenge—maximizing durability while minimizing mass—will apply to the next lunar vehicles, meaning that students will be working to solve this challenge for the moonbuggy race at the same time NASA engineers are working on similar problems.

Student innovators from the University of Evansville, in Evansville, Indiana, sped past 23 teams from around the globe to win the college division of the race. The Evansville team posted the day's fastest race time, completing the harrowing course in just 4 minutes and 25 seconds. Finishing in the top three along with Evansville were second-place winners from Murray State University, in Murray, Kentucky, and third-place racers representing Canada's Carleton University, in Ottawa, Ontario. ❖

Cold War from history books, plan to launch a satellite to commemorate the Nation's first successful satellite.

MSU students from a variety of disciplines have been building a namesake of Explorer-1 for nearly 3 years. They hope to launch the Explorer-1 (Prime) in December 2008.

Explorer-1 (Prime) is being built by the Space Science and Engineering Lab and its students for the Montana Space Grant Consortium. The satellite—an aluminum cube that measures about 4 inches per side—will hold instruments to detect radiation and a power supply to run those instruments. It will also contain one of the original Van Allen Geiger tubes that Dr. James Van Allen provided to MSU a few months before he died.

Once the satellite is completed, it will be sent to California Polytechnic State University to be placed in a container with two other satellites the same size. The container will then be mounted on a rocket and launched into space. Ham radio operators from all over the world should be able to track it during its 4-month orbit about 440 miles above them, making one orbit around the Earth every 90 minutes.

Partnership News

NASA cultivates partnerships with private industry, academia, and other government agencies to bring its science back down to Earth. By contributing time, facilities, and a wealth of technical expertise, NASA enriches the lives of people everywhere with these successful partnerships.

Partnership News

The Innovative Partnerships Program aims to provide leveraged technology for NASA's mission directorates, programs, and projects through investments and technology partnerships with industry, academia, government agencies, and national laboratories. The following stories highlight some of the exceptional results of these many partnerships.

Three NASA Technologies Inducted into the Space Technology Hall of Fame

On April 10, 2008, three NASA technologies were inducted into the Space Foundation's Space Technology Hall of Fame. Jet Propulsion Laboratory (JPL) received the honor for application of its imaging software to medical uses. Kennedy Space Center was honored for its role in the development of the ResQPOD, a noninvasive circulation-enhancing device, and was commended for the role it played in developing the microspheres now found in the Petroleum Remediation Product (PRP).

The Space Foundation, in cooperation with NASA, established the Space Technology Hall of Fame in 1988 to increase public awareness of the benefits resulting from space exploration programs and to encourage further innovation. Each year the Space Foundation recognizes unique and valuable products that originated with space technology.

Now celebrating its 20th anniversary, the Space Technology Hall of Fame has inducted 56 technologies to date, as well as honored the organizations and individuals who transformed space technology into commercial products that improve the quality of life for all humanity.

The first 2008 inductee, ArterioVision software is a diagnostic tool that is used in conjunction with a standard ultrasound to precisely measure the thickness of the two inner layers of the carotid artery. Arterial thickening can provide the earliest evidence of atherosclerosis, or hardening of the arteries. Initially developed at JPL through its Innovative Partnerships Program, this technology is derived from the Video Imaging Communication and Retrieval software used to process pictures from spacecraft imagery. ArterioVision allows doctors to measure the carotid intima media thickness (CIMT) to determine the age and health of a patient's arteries and better predict and prevent their risk for heart disease and stroke.

The ArterioVision CIMT procedure is approved by the U.S. Food and Drug Administration. Medical Technologies International Inc., of Palm Desert, California, the company that patented the software, was inducted along with JPL and the University of Southern California's Keck School of Medicine–Atherosclerosis Research Unit.

ResQPOD is a noninvasive medical device that improves cardiac output and blood flow to the brain compared to conventional resuscitation techniques. Developed through a collaborative research effort between Kennedy, the U.S. Army, and private industry, this device is used to help astronauts reacquaint with the feeling of gravity by quickly and effectively increasing the circulation of blood flow to the brain.

ResQPOD is used by emergency medical services and hospitals across the country for patients suffering orthostatic intolerance (breathing problems) and cardiac arrest or other conditions attributed to low blood pressure. It works by increasing blood flow to the heart and brain, which is critical to improving survival rates with normal neurological function, until the heart can be restarted. The U.S. military also uses ResQPOD on the battlefield to reduce intracranial pressure resulting from

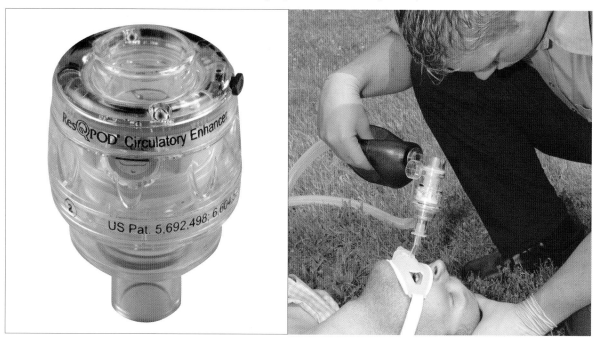

The ResQPOD Circulatory Enhancer increases blood flow to the heart and brain during CPR, which in turn increases the opportunity for survival and normal neurological outcome.

to preventing ground water contamination at petroleum storage facilities.

NASA Nanotubes Help Advance Brain Tumor Research

The potential of carbon nanotubes to diagnose and treat brain tumors is being explored through a partnership between JPL and City of Hope, a leading cancer research and treatment center in Duarte, California.

Nanotechnology may help revolutionize medicine in the future with its promise to play a role in selective cancer therapy. City of Hope researchers are working to boost the brain's own immune response against tumors by delivering cancer-fighting agents via nanotubes. A nanotube is about 50,000 times narrower than a human hair, but its length can extend up to several centimeters.

According to Dr. Behnam Badie, City of Hope's director of neurosurgery and of its brain tumor program, if nanotube technology can be effectively applied to brain tumors, it might also be used to treat stroke, trauma, neurodegenerative disorders, and other disease processes in the brain.

"I'm very optimistic of how this nanotechnology will work out," he said. "We are hoping to begin testing in humans in about 5 years, and we have ideas about where to go next."

The Nano and Micro Systems (NAMS) group at JPL, which has been researching nanotubes since about 2000, creates these tiny, cylindrical multiwalled carbon tubes for City of Hope.

City of Hope researchers, who began their quest in 2006, found good results: The nanotubes, which they used on mice, were nontoxic in brain cells, did not change cell reproduction and were capable of carrying DNA and siRNA, two types of molecules that encode genetic information.

JPL's NAMS group grows the nanotubes on silicon strips a few square millimeters in area. The growth process

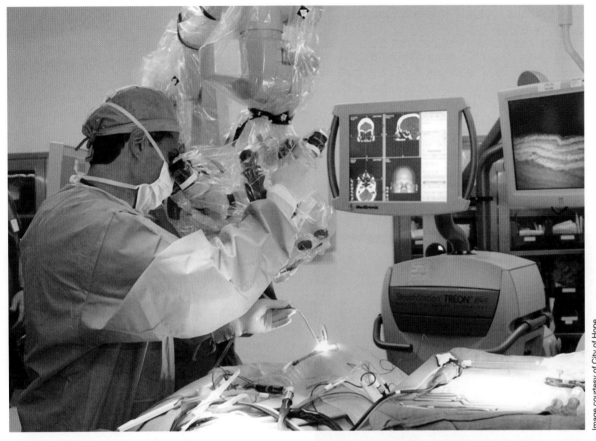

Image courtesy of City of Hope.

Dr. Behnam Badie, director of the Department of Neurosurgery and the brain tumor program at City of Hope, performs a minimally invasive procedure to surgically remove a pituitary tumor. Nanotube technology may help in the development of new treatments that would require only minimally invasive procedures, no matter the location of the brain tumor.

head trauma injuries. Advanced Circulatory Systems, of Minneapolis, Minnesota, and the Kennedy Biomedical Laboratory were inducted as the innovating organizations behind ResQPOD technology.

PRP is a technology that safely and permanently removes petroleum-based pollutants from water. The delivery system of this water treatment process grew out of NASA biological encapsulation research and experimentation in the orbital production of microspheres.

PRP uses microcapsules, tiny balls of beeswax with hollow centers, which absorb and bind with petroleum or other hydrocarbon products. The microspheres then serve as nutrients to assist naturally occurring microbes in soil or water to biodegrade contaminates. Universal Remediation Inc., of Pittsburgh, Pennsylvania, the inductee organization for PRP, has developed a number of customized products using PRP technology to treat environmental contaminants from small boating spills forms them into hollow tubes as if by rolling sheets of graphite-like carbon.

Carbon nanotubes are extremely strong, flexible, heat-resistant, and have very sharp tips. Consequently, JPL uses nanotubes as field-emission cathodes—vehicles that help produce electrons—for various space applications such as X-ray and mass spectroscopy instruments, vacuum microelectronics, and high-frequency communications.

"Nanotubes are important for miniaturizing spectroscopic instruments for space applications, developing extreme environment electronics, as well as for remote sensing," said Harish Manohara, the technical group supervisor for the NAMS group.

Nanotubes are a fairly new innovation, so they are not yet routinely used in current NASA missions, he added. However, they may be used in gas analysis or mineralogical instruments for future missions to Mars, Venus, and the Jupiter system.

JPL's collaboration with City of Hope began last year, after Manohara, Badie, and Dr. Babak Kateb, City of Hope's former director of research and development in the brain tumor program, discussed using nanostructures to better diagnose and treat brain cancer. Badie said his team's nanomedical research continues, and the next goal will be to functionalize and attach inhibitory RNA to the nanotubes and deliver it to specific areas of the brain.

The JPL and City of Hope teams published the results of the study earlier this year in the journal NeuroImage.

Badie says that JPL's contribution to City of Hope's nanomedicine research has been invaluable.

"The fact that we can get pristine and really clean nanotubes from Manohara's department is unique," he said. "The fact that we are both collaborating for biological purposes is also really unique."

The collaboration between JPL and City of Hope is conducted under NASA's Innovative Partnerships Program, designed to bring benefits of the space program to the public.

NASA Unveils Cosmic Images Book in Braille for Blind

At a ceremony held in January 2008, at the National Federation of the Blind, NASA unveiled a new book that brings majestic images taken by its Great Observatories to the fingertips of the blind. The Great Observatories include NASA's Hubble, Chandra, and Spitzer space telescopes.

"Touch the Invisible Sky" is a 60-page book with color images of nebulae, stars, galaxies, and some of the telescopes that captured the original pictures. Each image is embossed with lines, bumps, and other textures. These raised patterns translate colors, shapes, and other

This is a composite image of N49, the brightest supernova remnant in optical light in the Large Magellanic Cloud. Data from Spitzer, Hubble, and Chandra contributed to this image.

intricate details of the cosmic objects, allowing visually impaired people to experience them. Braille and large-print descriptions accompany each of the book's 28 photographs, making the book's design accessible to readers of all visual abilities.

The celestial objects are presented as they appear through visible-light telescopes and different spectral regions invisible to the naked eye, from radio to infrared, visible, ultraviolet and X-ray light. The book introduces the concept of light and the spectrum and explains how the different observatories complement each others' findings. Readers take a cosmic journey beginning with images of the Sun, and travel out into the galaxy to visit relics of exploding and dying stars, as well as the Whirlpool galaxy and colliding Antennae galaxies.

"Touch the Invisible Sky" was written by astronomy educator and accessibility specialist Noreen Grice of You Can Do Astronomy LLC and the Museum of Science, Boston, with authors Simon Steel, an astronomer with the Harvard-Smithsonian Center for Astrophysics, in Cambridge, Massachusetts, and Doris Daou, an astronomer at NASA Headquarters.

"About 10 million visually impaired people live in the United States," Grice said. "I hope this book will be a unique resource for people who are sighted or blind to better understand the part of the universe that is invisible to all of us."

The book will be available to the public through a wide variety of sources, including the National Federation of the Blind, Library of Congress repositories, schools for the blind, libraries, museums, science centers, and Ozone Publishing.

"We wanted to show that the beauty and complexity of the universe goes far beyond what we can see with our eyes!" Daou said.

"The study of the universe is a detective story, a cosmic 'CSI,' where clues to the inner workings of the universe are revealed by the amazing technology of modern telescopes," Steel said. "This book invites everyone to join in the quest to unlock the secrets of the cosmos."

"One of the greatest challenges faced by blind students who are interested in scientific study is that certain kinds of information are not available to them in a nonvisual form," said Marc Maurer, president of the National Federation of the Blind. "Books like this one are an invaluable resource because they allow the blind access to information that is normally presented through visual observation and media. Given access to this information, blind students can study and compete in scientific fields as well as their sighted peers."

The prototype for this book was funded by an education grant from the Chandra mission, and production was a collaborative effort by the NASA space science missions, which provided the images, and other Agency sources.

NASA's Advanced Technology Peers Deep Inside Hurricanes

Determined to understand why some storms grow into hurricanes while others fizzle, NASA scientists recently looked deep into thunderstorms off the African coast using satellites and airplanes.

A team of international scientists, including NASA researchers, journeyed to the west coast of Africa. Their mission was to better understand why some clusters of thunderstorms that drift off the African coast, known as easterly waves, develop into furious hurricanes, while others simply fade away within hours.

A major component of the campaign, called the NASA African Monsoon Multidisciplinary Analyses (NAMMA), was to study the Saharan Air Layer. The layer is a mass of very dry, dusty air that forms over the Sahara Desert and influences the development of tropical cyclones, the general name given to tropical depressions, storms, and hurricanes.

This mission was unique, because it incorporated NASA's state-of-the-art technology in space and in the

Hurricanes require a special set of conditions, including ample heat and moisture, conditions that exist primarily over warm tropical oceans.

air. With sophisticated satellite data and aircraft, scientists are better able to examine the tug-of-war between forces favorable for hurricane development—warm sea surface temperatures and rotating clusters of strong thunderstorms—and forces that suppress hurricanes such as dust particles and changing wind speed and direction at high altitudes.

"Most late-season Atlantic basin hurricanes develop from African easterly waves, so improving our knowledge of these hurricane seedlings is critical," said Ramesh Kakar, program manager for NAMMA at NASA Headquarters. "Several studies have shown that the Saharan Air Layer suppresses hurricane development, but the exact

mechanisms are very unclear, and it remains a wild card in the list of ingredients necessary for hurricane formation."

NASA's Moderate Resolution Imaging Spectro-radiometer instrument on the Terra and Aqua satellites identified the location, size, and intensity of dust plumes throughout the mission. Using other satellites, scientists could then determine any possible connection between dust outbreaks and changes in tropical easterly waves. The Tropical Rainfall Measuring Mission satellite, for instance, provided information on rainfall and thunderclouds, while the Quick Scatterometer satellite identified how low-level winds were rotating, both critical elements in hurricane formation.

NASA scientists also used a satellite product designed specifically to assess the strength of the Saharan Air Layer that uses imagery from Meteosat, a European satellite. Well-developed regions of the Saharan Air Layer were easily identified by measuring tiny dust particles and atmospheric water vapor content. Multiple images taken over time tracked dust movement and evolution across the Atlantic.

After analyzing satellite data, researchers flew aircraft into specific, targeted areas to probe storm clouds over a very short time and small area to learn how microscopic dust particles, called aerosols, interact with cloud droplets contained in thunderstorms. Aerosols potentially influence rainfall and the overall structure and future strength of a developing tropical cyclone. The extreme dry air, warm temperatures, and wind shear within these elevated dust layers may also weaken fledgling tropical cyclones.

Scientists flew a total of 13 aircraft missions inside 7 storm systems. NASA's DC-8 research aircraft contained numerous instruments to take measurements deep inside clouds, the environment of thunderstorms, and the Saharan Air Layer. Researchers also took advantage of several aircraft probes and especially dropsondes, a sensor attached to a parachute that is dropped into storm clouds. It typically collects data on wind speed and direction, temperature, humidity, and pressure that are relayed to a computer in the airplane.

Aircraft sensors and laser devices called lidars measured water vapor content and cloud, dust, and precipitation particle sizes, shapes, and types. Revolutionary radar on the aircraft was also used to gather better details on the intensity of rainfall and where exactly it was falling.

One special sensor aboard the DC-8, called the High-Altitude MMIC Sounding Radiometer, provided a 3-D distribution of temperature and water vapor in the atmosphere. The sensor is ideal for hurricane studies, since it can look through thick clouds and probe into the interior of the storms. It has also led to the development of a new microwave sounder for geostationary satellites, GeoSTAR,

which will make it possible to monitor the interior of hurricanes continuously without having to wait for a satellite to pass overhead.

Throughout the field mission, a Web-based real-time mission monitor, developed by Marshall, allowed scientists to track the progress of the experiment from anywhere on the globe using a standard Internet connection.

"Through the use of sophisticated technology, NAMMA provided an excellent opportunity to advance our understanding of tropical cyclones, as we gathered data on the critical elements at both the very small and large scales, from microscopic dust to air currents spanning hundreds of miles," said Jeff Halverson, one of four NAMMA mission scientists. "Much of the data gathered is still being analyzed, but the preliminary findings are very promising."

NASA scientists will also compare NAMMA findings to data from previous missions that took place in the Caribbean and Gulf of Mexico. The results should help determine the role of factors universal to hurricane formation and those uniquely dependent on location.

Breakthrough Map of Antarctica Lays Groundwork for New Discoveries

A team of researchers from NASA, the U.S. Geological Survey (USGS), the National Science Foundation, and the British Antarctic Survey unveiled a newly completed map of Antarctica that is expected to revolutionize research of the continent's frozen landscape.

The Landsat Image Mosaic of Antarctica is a result of NASA's state-of-the-art satellite technologies and an example of the prominent role NASA continues to play as a world leader in the development and flight of Earth-observing satellites.

The map is a realistic, nearly cloudless satellite view of the continent at a resolution 10 times greater than ever before with images captured by the NASA-built Landsat 7. With the unprecedented ability to see features half the size of a basketball court, the mosaic offers the

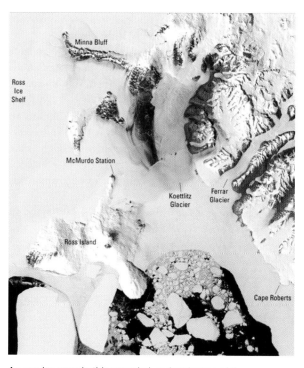

As can be seen in this sample Landsat image of the area around McMurdo Station, the new mosaic reveals (in unprecedented detail) the ice shelves, mountains, and glaciers that make Antarctica a fascinating and important place to study.

most geographically accurate, true-color, high-resolution views of Antarctica to date.

"This mosaic of images opens up a window to the Antarctic that we just haven't had before," said Robert Bindschadler, chief scientist of the Hydrospheric and Biospheric Sciences Laboratory at Goddard Space Flight Center. "It will open new windows of opportunity for scientific research as well as enable the public to become much more familiar with Antarctica and how scientists use imagery in their research. This innovation is like watching high-definition TV in living color versus watching the picture on a grainy black-and-white television. These scenes don't just give us a snapshot, they provide a time-lapse historical record of how Antarctica has changed and will enable us to continue to watch changes unfold."

Researchers can use the detailed map to better plan scientific expeditions. The mosaic's higher resolution gives researchers a clearer view over most of the continent to help interpret changes in land elevation in hard-to-access areas. Scientists also think the true-color mosaic will help geologists better map various rock formations and types.

To construct the new Antarctic map, researchers pieced together more than a thousand images from 3 years of Landsat observations. The resulting mosaic gives researchers and the public a new way to explore Antarctica through a free, public-access Web portal. Eight different versions of the full mosaic are available to download.

In 1972, the first satellite images of the Antarctic became available with the launch of NASA's Earth Resources Technology Satellite (later renamed Landsat). The series of Landsat satellites have provided the longest, continuous global record of land surface and its historical changes in existence. Prior to these satellite views, researchers had to rely on airplanes and survey ships to map Antarctica's ice-covered terrain.

Images from the Landsat program, now managed by the USGS, led to more precise and efficient research results as the resolution of digital images improved over the years with upgraded instruments on each new Earth-observing satellite.

"We have significantly improved our ability to extract useful information from satellites as embodied in this Antarctic mosaic project," said Ray Byrnes, liaison for satellite missions at the USGS, in Reston, Virginia. "As technology progressed, so have the satellites and their image resolution capability. The first three in the Landsat series were limited in comparison to Landsats 4, 5, and 7."

Bindschadler, who conceived the project, initiated NASA's collection of images of Antarctica for the mosaic project in 1999. He and NASA colleagues selected the images that make up the mosaic and developed new techniques to interpret the image data tailored to the project.

The mosaic is made up of about 1,100 images from Landsat 7, nearly all of which were captured between 1999 and 2001. The collage contains almost no gaps in the landscape, other than a doughnut hole-shaped area at the South Pole, and shows virtually no seams.

"The mosaic represents an important U.S.-U.K. collaboration and is a major contribution to the International Polar Year," said Andrew Fleming of British Antarctic Survey, in Cambridge, England. "Over 60,000 scientists are involved in the global International Polar Year initiative to understand our world. I have no doubt that polar researchers will find this mosaic, one of the first outcomes of that initiative, invaluable for planning science campaigns."

NASA has 14 Earth-observing satellites in orbit with activities that have direct benefit to humankind. After NASA develops and tests new technologies, the Agency transfers activities to other Federal agencies for vital meteorology and climate satellite services. The satellites have helped revolutionize the information that emergency officials have to respond to natural disasters like hurricanes and wildfires.

NASA Climate Change 'Peacemakers' Aided Nobel Effort

It's not every day that a NASA scientist can wake up and think, "Hey, I did something for world peace." But on Monday, December 10, 2007, many NASA Earth scientists did exactly that.

In Oslo, Norway, the King of Sweden presented the shared 2007 Nobel Peace Prize to former U.S. Vice President Al Gore and to representatives of a United Nations panel that has spent two decades assessing Earth's changing climate and predicting where it is headed. Hundreds of NASA scientists, including some from JPL, contributed to the United Nations effort, working with thousands of their colleagues from more than 150 countries.

Announcing the Nobel Peace Prize, the Norwegian Nobel Committee said the scientific reports issued since 1990 by the United Nations Intergovernmental Panel on Climate Change (IPCC) have "created an ever-broader informed consensus about the connection between human activities and global warming." The peacemaking value of this scientific finding, according to the committee, is that human-induced changes in climate may cause "large-scale migration and lead to greater competition for the Earth's resources" and an "increased danger of violent conflicts and wars."

The Fourth IPCC Assessment, released this year in four reports, presented the strongest findings thus far that human activities are altering Earth's climate and that the impacts of climate change are occurring already.

The 2007 Nobel Peace Prize honored the work of climate scientists around the world, including many at NASA.

When the First IPCC Assessment was reported in 1990, NASA was building on a history of Earth remote sensing, developing and preparing to deploy the Earth Observing Satellite system to determine the extent, causes, and regional consequences of global climate change. In the recent Fourth Assessment, scientists were informed by more than 8 years of systematic, global observations of the Earth system. Satellite measurements have revealed fundamental changes in Earth's climate, including temperatures and rainfall, ice extent and properties, and sea levels, as well as physical, chemical, and ecological impacts of climate change. NASA satellite measurements contributed immeasurably to enable the IPCC's strongest conclusions thus far.

"NASA is best known for its cutting-edge satellite instruments and global measurements of Earth from space, but we contribute a lot more than that to climate change science," said Michael Gunson, acting chief scientist in JPL's Earth Science and Technology Directorate. "NASA's role extends far beyond space-based measurements into the research to build our understanding of climate change, enabling the critical work of the IPCC."

NASA instruments, data, analysis, and modeling all contributed to the bedrock of the IPCC report: the hundreds of papers published each year in scientific journals, many authored by NASA scientists and many others using NASA observations. The authors of the report draw on this ever-growing body of new knowledge to form their conclusions about climate change.

"The most remarkable thing about the process of assembling an IPCC report is that you can actually get thousands of independent-minded and critical scientists to work together without killing each other," said Bruce Wielicki, senior scientist for Earth science at Langley Research Center.

Wielicki contributed a portion of a chapter in the latest science assessment on how Earth's "energy budget," the ebb and flow of radiant energy from the Sun and our planet, has changed as measured by satellites. He began the project in October 2004 and, working with a team of 10 scientists, completed a compact summary of the latest research on the topic 20 months later. Like each section of the IPCC reports, Wielicki's section went through repeated rounds of critiques by other scientists.

NASA's Cynthia Rosenzweig, a plant and soil scientist at the Goddard Institute for Space Studies, in New York, coordinated a key chapter in the new report on the impact of climate change—an effort that took 4 years. "There were many, many late nights as we worked under strict deadlines to draft the chapter and revise it based on thousands of comments from reviewers, each of which had to be documented and responded to," Rosenzweig recalls.

"But the toughest part of the entire effort was the last step: reviewing our final draft with government officials," Rosenzweig says. Before each IPCC report is published, the lead authors sit down with diplomats, lawyers, and environmental officials from around the world to review their findings, page by page. "These week-long meetings are very challenging as you respond to all sorts of concerns and questions. But this process is the real beauty of the IPCC. The final documents that emerge represent a consensus view of the world's scientific community and government delegates."

The Nobel-winning IPCC reports are unparalleled as the most authoritative source of climate science, says Wielicki. "When I give public lectures on climate, I tell my audience that there are three laws of solid information on climate change: IPCC, IPCC, and IPCC."

The IPCC effort has also boosted public awareness of this critical area of science. "By collecting together the current scientific thinking on climate change, the IPCC showed the world the value of the type of science we are doing at NASA," said JPL's Gunson. "And that has really engaged the public, many of whom were surprised that NASA does climate research. It has really motivated a new interest in the work we do here day in and day out."

NASA Aircraft Aids Southern California Firefighting Effort

In response to a request from the California Office of Emergency Services and the National Interagency Fire Center, NASA flew an aircraft equipped with sophisticated infrared imaging equipment to assist firefighters battling several Southern California wildfires.

The Ikhana unmanned aircraft system, a Predator B, modified for civil science and research missions, was launched from its base at Dryden Flight Research Center at Edwards Air Force Base. It then flew over the major blazes burning in the Lake Arrowhead and Running Springs areas and then down into San Diego County to image wildfires raging in that area. The aircraft was controlled remotely by pilots in a ground control station at Dryden.

The Ikhana carried the Autonomous Modular Scanner, a thermal-infrared imaging system developed at Ames Research Center in Northern California. The system is capable of peering through heavy smoke and darkness to see hot spots, flames, and temperature differences, processing the imagery onboard, and then transmitting that information in near real time so it can aid fire incident commanders in allocating their firefighting resources.

The images were transmitted through a communications satellite to Ames where the imagery was placed on an Ames Web site, combined with Google Earth maps, and then transmitted to the interagency fire center in Boise, Idaho, where it was then made available to incident commanders in the field.

The system was validated recently during a series of wildfire imaging demonstration missions conducted by NASA and the U.S. Forest Service in August and September 2007.

Each flight was coordinated with the FAA to allow the remotely piloted aircraft to fly within the national airspace while maintaining separation from other aircraft.

NASA Tsunami Research Makes Waves in Science Community

A wave of new NASA research on tsunamis has yielded an innovative method to improve existing tsunami warning systems, and a potentially groundbreaking new theory on the source of the December 2004 Indian Ocean tsunami.

In one study, published last fall in Geophysical Research Letters, researcher Y. Tony Song, of JPL, demonstrated that real-time data from NASA's network of Global Positioning System (GPS) stations can detect ground motions preceding tsunamis and reliably estimate a tsunami's destructive potential within minutes, well before it reaches coastal areas. The method could lead to development of more reliable global tsunami warning systems, saving lives and reducing false alarms.

Conventional tsunami warning systems rely on estimates of an earthquake's magnitude to determine whether a large tsunami will be generated. Earthquake magnitude is not always a reliable indicator of tsunami potential, however. The 2004 Indian Ocean quake generated a huge tsunami, while the 2005 Nias (Indonesia) quake did not, even though both had almost the same

Ikhana carries the Autonomous Modular Scanner payload developed by NASA's Ames Research Center, equipment that incorporates a sophisticated imaging sensor and real-time data communications equipment. The sensor is capable of peering through thick smoke and haze to record hot spots and the progression of wildfires over a lengthy period.

Thermal-infrared imaging sensors on Ikhana recorded this image of the Harris Fire in San Diego County on October 24, 2007, with hot spots along the ridgeline in left center clearly visible.

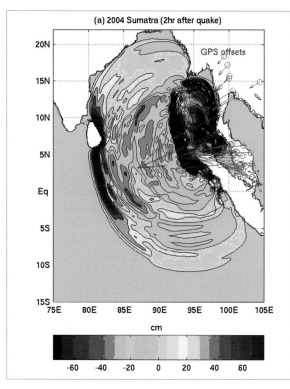

(a) 2004 Sumatra (2hr after quake)

Using GPS data (purple arrows) to measure ground displacements, scientists replicated the December 2004 Indian Ocean tsunami, whose crests and troughs are shown here in reds and blues, respectively. The research showed GPS data can be used to reliably estimate a tsunami's destructive potential within minutes.

magnitude from initial estimates. Between 2005 and 2007, five false tsunami alarms were issued worldwide. Such alarms have negative societal and economic effects.

Song's method estimates the energy an undersea earthquake transfers to the ocean to generate a tsunami by using data from coastal GPS stations near the epicenter. With these data, ocean floor displacements caused by the earthquake can be inferred. Tsunamis typically originate at undersea boundaries of tectonic plates near the edges of continents.

"Tsunamis can travel as fast as jet planes, so rapid assessment following quakes is vital to mitigate their hazard," said Ichiro Fukumori, a JPL oceanographer not involved in the study. "Song and his colleagues have demonstrated that GPS technology can help improve both the speed and accuracy of such analyses."

Song's method works as follows: an earthquake's epicenter is located using seismometer data. GPS displacement data from stations near the epicenter are then gathered to derive seafloor motions. Based upon these data, local topography data, and new theoretical developments, a new "tsunami scale" measurement from 1 to 10 is generated, much like the Richter scale used for earthquakes. Song proposes using the scale to make a distinction between earthquakes capable of generating destructive tsunamis from those unlikely to do so.

To demonstrate his methodology on real earthquake-tsunamis, Song examined three historical tsunamis with well-documented ground motion measurements and tsunami observations: Alaska in 1964; the Indian Ocean in 2004; and Nias Island, Indonesia in 2005. His method successfully replicated all three. The data compared favorably with conventional seismic solutions that usually take hours or days to calculate.

Song said many coastal GPS stations are already in operation, measuring ground motions near earthquake faults in real time once every few seconds. "A coastal GPS network established and combined with the existing International GPS Service global sites could provide a more reliable global tsunami warning system than those available today," he said.

The theory behind the GPS study was published in the December 20, 2007 issue of Ocean Modelling. Song and his team from JPL; the California Institute of Technology, in Pasadena; the University of California, Santa Barbara; and Ohio State University, in Columbus, theorized most of the height and energy generated by the 2004 Indian Ocean tsunami resulted from horizontal, not vertical, faulting motions. The study uses a 3-D earthquake-

tsunami model based on seismograph and GPS data to explain how the fault's horizontal motions might be the major cause of the tsunami's genesis.

Scientists have long believed tsunamis form from vertical deformation of seafloor during undersea earthquakes. However, seismograph and GPS data show such deformation from the 2004 Sumatra earthquake was too small to generate the powerful tsunami that ensued. Song's team found horizontal forces were responsible for two-thirds of the tsunami's height, as observed by three satellites (NASA's Jason, the U.S. Navy's Geosat Follow-on, and the European Space Agency's Environmental Satellite), and generated five times more energy than the earthquake's vertical displacements. The horizontal forces also best explain the way the tsunami spread out across the Indian Ocean. The same mechanism was also found to explain the data observed from the 2005 Nias earthquake and tsunami.

Coauthor C.K. Shum, of Ohio State University, said the study suggests horizontal faulting motions play a much more important role in tsunami generation than previously believed. "If this is found to be true for other tsunamis, we may have to revise some early views on how tsunamis are formed and where mega tsunamis are likely to happen in the future," he said.

Centennial Challenge Winner Delivers First Commercial Space Suit Gloves

NASA's Centennial Challenges program of prize contests stimulate innovation and competition in solar system exploration and ongoing NASA mission areas. By making awards based on actual achievements, instead of proposals, Centennial Challenges seeks novel solutions to NASA's mission challenges from non-traditional sources of innovation in academia, industry, and the public.

One of seven such challenges, the Astronaut Glove Challenge is designed to promote the development of glove joint technology, resulting in a highly dexterous and flexible glove that can be used by astronauts over long periods of time for space or planetary surface excursions.

Flagsuit LLC, a new startup founded by NASA Astronaut Glove Challenge winner Peter Homer, shipped its first commercially produced space suit gloves to Los Angeles-based Orbital Outfitters in February 2008 under a joint development agreement.

The gloves are designed to be used with the Industrial Suborbital Space Suit-Crew (IS3C), which was unveiled by Orbital Outfitters in October 2007. The gloves will be used for integrated suit testing and evaluation, and feature a patent-pending joint design that makes the fingers more flexible under pressure, increasing dexterity while reducing hand fatigue. The gloves are manufactured using a new process that eliminates time-consuming adjustments to adapt the fit to the wearer's hands, producing a ready-to-wear garment that literally "fits like a glove." Flagsuit is currently implementing a preliminary production capability with support from the Maine Technology Institute. The commercial space suit gloves are direct descendants of the design that won the 2007 NASA challenge. "We haven't lost sight of how we got started," said Flagsuit

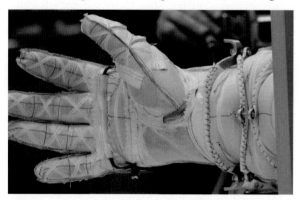

Peter Homer of Southwest Harbor, Maine, won $200,000 from NASA for his entry in the Astronaut Glove Challenge. He has since founded a company, Flagsuit LLC, to manufacture the gloves for future space missions.

founder Peter Homer. "The goal of the NASA Centennial Challenges is to spur development of needed technologies by using prize competitions to encourage private investment. Through our work on the commercial side of the market, Flagsuit is advancing innovations, such as 'Made to Fit,' up the technology readiness curve at no cost to NASA. Combined with our flexible joint technology, this will bring tremendous value to NASA as well as to our commercial partners." Flagsuit and Orbital Outfitters will continue to work together to refine the design of gloves and other space suit elements.

NASA and India Sign Agreement for Future Cooperation

At a ceremony at Kennedy's visitor complex, NASA Administrator Michael Griffin and Indian Space Research Organization Chairman G. Madhavan Nair signed a framework agreement establishing the terms for future cooperation between the two agencies in the exploration and use of outer space for peaceful purposes.

"I am honored to sign this agreement with the Indian Space Research Organization," Griffin said. "This agreement will allow us to cooperate effectively on a wide range of programs of mutual interest. India has extensive space-related experience, capabilities and infrastructure, and will continue to be a welcome partner in NASA's future space exploration activities."

According to the framework of the agreement, the two agencies will identify areas of mutual interest and seek to develop cooperative programs or projects in Earth and space science, exploration, human space flight, and other activities. The agreement replaces a previous agreement signed in December 1997, which fostered bilateral cooperation in the areas of Earth and atmospheric sciences.

In addition to a long history of cooperation in Earth science, NASA and the Indian Space Research Organization also are cooperating on India's first mission to the Moon, Chandrayaan-1, which will be launched later

this year. NASA is providing 2 of the 11 instruments on the spacecraft: the Moon mineralogy mapper instrument and the miniature synthetic aperture radar instrument.

NASA Data Link Pollution to Rainy Summer Days in the Southeast

Rainfall data from a NASA satellite show that summertime storms in the Southeastern United States shed more rainfall midweek than on weekends. Scientists say air pollution from humans is likely driving that trend.

The link between rainfall and the day of the week is evident in data from NASA's Tropical Rainfall Measuring Mission satellite, known as TRMM. Midweek storms tend to be stronger, drop more rain, and span a larger area across the Southeast compared to calmer and drier weekends. The findings are from a study led by Thomas Bell, an atmospheric scientist at Goddard Space Flight Center. Bell said the trend could be attributed to atmospheric pollution from humans, which also peaks midweek.

"It's eerie to think that we're affecting the weather," said Bell, lead author of the study published in the American Geophysical Union's Journal of Geophysical Research. "It appears that we're making storms more violent."

Rainfall measurements collected from ground-based gauges can vary from one gauge site to the next because of fickle weather patterns. So, to identify any kind of significant weekly rainfall trend, Bell and colleagues looked at the big picture from Earth's orbit. The team collected data from instruments on the TRMM satellite, which they used to estimate daily summertime rainfall averages from 1998 to 2005 across the entire Southeast.

The team found that, on average, it rained more between Tuesday and Thursday than from Saturday through Monday. Newly analyzed satellite data show that summer 2007 echoed the midweek trend with peak rainfall occurring late on Thursdays. In addition,

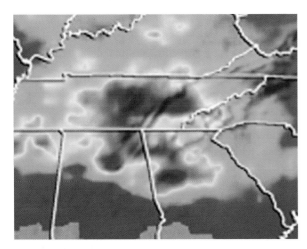

Torrential rainfall from a 2003 storm in the Southeast resulted in massive accumulations of rain (red). Similar data from NASA's TRMM satellite has revealed that more rain falls midweek.

midweek increases in rainfall were more significant in the afternoon, when the conditions for summertime storms are in place. Based on satellite data, afternoon rainfall peaked on Tuesdays, with 1.8 times more rainfall than on Saturdays, which experienced the least amount of afternoon rain.

The team used ground-based data, along with vertical wind speed and cloud height measurements, to help confirm the weekly trend in rainfall observed from space.

To find out if pollution from humans indeed could be responsible for the midweek boost in rainfall, the team analyzed particulate matter, the concentrations of airborne particles associated with pollution, across the U.S. from 1998 to 2005. The data, obtained from the U.S. Environmental Protection Agency, showed that pollution tended to peak midweek, mirroring the trend observed in the rainfall data.

"If two things happen at the same time, it doesn't mean one caused the other," Bell said. "But it's well known that particulate matter has the potential to affect how clouds behave, and this kind of evidence makes the argument stronger for a link between pollution and heavier rainfall."

Scientists long have questioned the effect of workweek pollution, such as emissions from traffic, businesses, and factories, on weekly weather patterns. Researchers know clouds are "seeded" by particulate matter. Water and ice in clouds grab hold around the particles, forming additional water droplets. Some researchers think increased pollution thwarts rainfall by dispersing the same amount of water over more seeds, preventing them from growing large enough to fall as rain. Still, other studies suggest some factors can override this dispersion effect.

In the Southeast, summertime conditions for large, frequent storms are already in place, a factor that overrides the rain-thwarting dispersion effect. When conditions are poised to form big storms, updrafts carry the smaller, pollution-seeded raindrops high into the atmosphere where they condense and freeze.

"It's the freezing process that gives the storm an extra kick, causing it to grow larger and climb higher into the atmosphere," Bell said. He and his colleagues found that the radar on the TRMM satellite showed that storms climb to high altitudes more often during the middle of the week than on weekends. These invigorated midweek storms, fueled by workweek pollution, could drop measurably more rainfall.

The trend doesn't mean it will always rain on weekday afternoons during summertime in the Southeast. Rather, "it's a tendency," according to Bell. But with the help of satellites, new insights into pollution's effect on weather one day could help improve the accuracy of rainfall forecasts, which Bell said, "probably under-predict rain during the week and over-predict rain on weekends."

SOFIA Completes Closed-Door Test Flights

NASA's Stratospheric Observatory for Infrared Astronomy, or SOFIA, has passed a significant mission milestone. The observatory has completed the first phase of experimental flight tests in a joint program by NASA and Deutsche Zentrum für Luft- und Raumfahrt (German Aerospace Center (DLR)). SOFIA's science and mission operations are managed jointly by the Universities Space Research Association (USRA) and the Deutsches SOFIA Institut (DSI). Tests confirmed the structural integrity and performance of the modified 747SP SOFIA aircraft that carries a huge infrared telescope.

The telescope measures nearly 10 feet in width and weighs almost 19 tons. It peers through a 16-foot-high door cut into SOFIA's 747 fuselage. During this test series, the aircraft flew five times with this external door closed. These flights tested the limits of the aircraft's capabilities in many areas, including aerodynamics, structural integrity, stability and control, and handling qualities.

"SOFIA is already a technological marvel, and will soon be a powerful tool for studying the birth and evolution of planets, stars, and galaxies," said Alan Stern, associate administrator of NASA's Science Mission Directorate. "The completion of its closed door testing phase is a major milestone on the way to SOFIA's inaugural science flights next year."

The SOFIA program also checked the functionality of the aircraft's cutting edge, German-built telescope. Engineers tested the ability of the instrument's control system to maintain its precise position when tracking a celestial object, even while the aircraft moves and maneuvers through the sky.

"The project finished a very important milestone on the path to the first astronomy work with the telescope, which is expected in early 2009," said Robert Meyer, SOFIA program manager at Dryden.

The aircraft will now undergo installation and integration of the remaining elements of the observatory before open-door test flights, scheduled to begin in late 2008. After completing the initial open-door test flight, limited science observation flights will begin in 2009. The science community will survey the universe with five specialized instruments on SOFIA as the observatory begins normal science observation flights in 2011. The observatory reaches full operational capabilities in 2014.

The SOFIA aircraft is based at Dryden's newly established Aircraft Operations Facility, in Palmdale, California, where it will remain for additional development, flight testing, and science flight operations. Dryden manages the SOFIA program and Ames manages the science project.

NASA Selects 302 Small Business Research and Technology Projects

NASA has awarded contracts to 302 small business proposals that address critical research and technology needs for Agency programs and projects. The awards are part of NASA's Small Business Innovation Research program (SBIR), and the Small Business Technology Transfer program (STTR).

The SBIR program selected 276 proposals for negotiation of Phase I contracts, and the STTR program chose 26 proposals for negotiation of Phase I contract awards. The selected SBIR projects have a total value of approximately $27.6 million. The selected STTR projects have a total value of approximately $2.6 million. The SBIR contracts will be awarded to 205 small, high-technology firms in 31 states. The STTR contracts will be awarded to 24 small, high-technology firms in 14 states. As part of the STTR program, the firms will partner with 22 universities and research institutions in 15 states.

SBIR and STTR are part of the Innovative Partnerships Program at NASA Headquarters, which works with U.S. industry to infuse pioneering technologies into NASA missions and transition them into commercially available products and services.

The SBIR program supports NASA's mission directorates by competitively selecting ventures that address specific technology gaps in mission programs and strives to complement other Agency research investments. Results from the program have benefited several NASA efforts, including air traffic control systems, Earth-

NASA's SOFIA infrared observatory and an F/A-18 safety chase during the first series of test flights to verify the flight performance of the modified Boeing 747SP.

observing spacecraft, the International Space Station, and the development of spacecraft for exploring the solar system.

Research topic areas among this group of selected proposals include:

- A simulated test bed for identifying dynamic air corridors to increase aircraft throughput

- Compact, 3-D scanning light detection and ranging for robotic navigation on the lunar surface, known as lidar

- Regenerative fuel cells for use on the lunar surface

- Ultra-high efficiency solar cells designed to operate on spacecraft in extreme environment missions

- High-efficiency transmitters for space communications that provide a significant improvement in its power output capability without an impact on the payload size and power

The SBIR program is a highly competitive, three-phase award system. It provides qualified small businesses—including women-owned and disadvantaged firms—with opportunities to propose unique ideas that meet specific research and development needs of the Federal government.

These contract awards are for Phase I, which is a feasibility study with as much as $100,000 in funding to evaluate the scientific and technical merit of an idea. The SBIR awards may last as long as 6 months. The STTR awards may last as long as 1 year. Phase II expands on the results on the development of Phase I; awards are for as much as $600,000 during as long as 2 years. Phase III is for the commercialization of the results of Phase II and requires the use of private sector or non-SBIR Federal funding.

Contractors submitted 1,500 Phase I SBIR proposals and 166 Phase I STTR proposals for competitive selection. The criteria used to choose the winning proposals included technical merit and feasibility; experience, qualifications, and facilities; effectiveness of the work plan; and commercial potential and feasibility.

Ames manages the program for the Innovative Partnerships Program office. NASA's 10 field centers manage the individual projects.

Inflatable Habitat Goes to McMurdo in Preparation for Moon, Mars

In September 2007, NASA, the National Science Foundation (NSF), and ILC Dover, of Frederica, Delaware, unveiled the Antarctic-bound inflatable habitat. Explorers residing at the NSF-managed McMurdo Station in the Antarctic will live in the inflatable structure through February 2009 and will report their experience. Sensors will also collect and report data. Testing the habitat in such an inhospitable setting will help scientists design similar habitats for use on the Moon or Mars.

"Testing the inflatable habitat in one of the harshest, most remote sites on Earth gives us the opportunity to see what it would be like to use for lunar exploration," said Paul Lockhart, director of Constellation Systems for NASA's Exploration Systems Mission Directorate.

By 2020, NASA hopes to return astronauts to the Moon, where they will set up a lunar outpost for long-duration stays; an inflatable habitat will provide protected living space while also being easily transportable to and across the lunar surface.

"Our habitation concepts have to be lightweight as well as durable," Lockhart said. "This prototype inflatable habitat . . . only takes four crew members a few hours to set up, permitting exploration beyond the initial landing area." Resembling a large inflatable tent, the structure is insulated and heated, has power, and is pressurized. It offers 384 square feet of living space.

Also interested in new habitat design is the NSF, which currently uses a bulky, complex, 50-year-old design known as a Jamesway hut, a version of the Quonset hut

A joint project among NASA, the National Space Foundation, and ILC Dover continues at the McMurdo Complex in Antarctica. Team members drill into the tundra to install a weather station adjacent to the inflatable habitat in the upper left portion of the image.

with an insulated fabric cover. In addition to studying the new data collected from the inflatable habitat's sensors, the NSF plans to study improvements in packing, transportation, power consumption, and damage tolerance.

Resembling an inflatable Quonset hut, the habitat is insulated and heated, has power, and offers 384 square feet of living space.

Ames GREEN Team and Google Connect on Environmental Research

The Global Research into Energy and the Environment at NASA (GREEN) symposium held at Ames Research Center on October 19, 2007, was the first in a NASA and Google Inc.-sponsored series which aims to create connections between NASA, academia, and the clean technology community.

The GREEN Team and Google next hosted a Director's Colloquium on December 5, 2007. Dr. Marty Hoofer, professor emeritus of physics and former chair of the Department of Applied Science at New York University, discussed breakthroughs in cost-effective space solar power—the beaming of high-intensity solar power from space at laser or microwave frequencies for electric power at the surface. The next day, experts from Ames, Glenn Research Center, and the University of California, Santa Cruz reviewed NASA's role in energy research and discussed how NASA might contribute to renewable energy research.

The third GREEN Team seminar was held on Tuesday, January 15, 2008, and included an overview of transportation policy and discussions of new approaches from industry and NASA, including not only new adaptations to bio-fuels such as sugar cane and methane, but also ideas for a rapid transportation network. Unimodal LLC's Chris Perkins presented his ideas on the Skytran, a small "car" that transports people and is powered by magnets.

Renowned environmental designer William McDonough spoke at Ames on February 5, 2008, about his "cradle to cradle" philosophy in the fourth GREEN Team seminar. Drawing inspiration from the effectiveness of natural systems, McDonough has been a pioneer in the sustainability movement since 1981; he helped design Google's new green campus at Ames in 2007, suggesting ways Google and Ames could have a more actively environmental space. In addition to using wind and solar power, McDonough suggested design ideas, such as constructing parking lots from permeable layers of stone and dirt to allow for better rainwater absorption, and shifting from merely "eco-efficient" actions like recycling to even more "eco-effective" strategies. Google consumes a great deal of electricity due to thousands of computer servers running simultaneously; because of this, many corporations considering "going green" are closely watching Google and NASA's collaboration.

The GREEN Team hosted another Director's Colloquium on March 11, 2008. Jim Woolsey, chair of the advisory boards of the Clean Fuels Foundation and the New Uses Council, connected security issues and climate change with possible solutions to American dependence on oil.

April's symposium addressed new ecosystem services, which include the management of environmental data from Earth-observing satellites and the use of supercomputers to model current climates. Ames Director S. Pete Worden stressed how the Google and NASA collaboration is addressing questions about stewardship of the environment.

NASA and the National Institutes of Health Partner for Health Research in Space

In September 2007, NASA and the National Institutes of Health (NIH) announced a partnership for health research in space. Joined by U.S. Senators Kay Bailey Hutchison, of Texas, and Barbara Mikulski, of Maryland, NASA Administrator Michael Griffin and NIH Director Elias A. Zerhouni signed the Memorandum of Understanding, the first of its kind between NASA and another government agency, which outlines how the U.S. segment of the International Space Station (ISS) will be used as a national research laboratory. The report also describes possible partnerships with other agencies and companies.

"I am extremely pleased that this collaborative effort is moving forward," said Zerhouni. "The station provides a unique environment where researchers can explore fundamental questions about human health issues—including how the body heals itself, fights infection, or develops diseases such as cancer or osteoporosis."

The agreement helps American scientists perform research aboard the ISS to answer questions about human health and diseases; NASA and the NIH expect it will help advance scientific discovery.

"The congressional designation as a national laboratory underscores the significance the American people place on the scientific potential of the space station," Griffin explained. "Not only will the station help in our efforts to explore the Moon, Mars, and beyond, its resources also can be applied for a much broader purpose—improving human health."

The facility at the station provides a virtually gravity-free environment where the cellular and molecular mechanisms that underlie human diseases can be explored. For example:

- The station provides a stable platform on which scientists can study the molecular basis for the effect of weightlessness on bone deterioration, which could benefit people who suffer from bone or muscle-wasting diseases.

- Studies of the brain and how it compensates for the absence of sensory input and spatial orientation may hold promise for people who suffer from balance disorders.

- Changes in human immunity during prolonged space travel could offer new hope to people with immune system issues.

- Remote health monitoring of astronauts may generate more technologies useful for Earth applications.

As part of the agreement, NIH and NASA will encourage space-related health research by exchanging information, providing expertise in areas of common interest, and sharing each other's research and development efforts. ❖

Innovative Partnerships Program

The Innovative Partnerships Program creates partnerships with industry, academia, and other sources to develop and transfer technology in support of national priorities and NASA's missions. The programs and activities resulting from the partnerships engage innovators and enterprises to fulfill NASA's mission needs and promote the potential of NASA technology.

Innovative Partnerships Program

The Innovative Partnerships Program (IPP) provides needed technology and capabilities to NASA's mission directorates, programs, and projects through investments and partnerships with industry, academia, government agencies, and national laboratories.

IPP has offices at each of NASA's 10 field centers, and elements that include: Technology Infusion, which manages the Small Business Innovation Research (SBIR) and Small Business Technology Transfer (STTR) programs and the IPP Seed Fund; the Innovation Incubator, which includes the Centennial Challenges and new efforts with the emerging commercial space sector; and Partnership Development, which includes intellectual property management and technology transfer. In 2007:

- IPP facilitated the Agency's entering into over 200 Space Act Agreements with private and other external entities for development of dual-use technology targeted to mission directorate technology needs.

- IPP provided $9.2 million in funding for 38 Seed Fund partnerships for development of a broad spectrum of technologies addressing specific mission directorate technology gaps. Partner and field center contributions of cash and in-kind resources will leverage these funds by a factor of four.

- IPP facilitated the signing of 35 license agreements and 598 Software Use Agreements.

- IPP facilitated the reporting of more than 1,200 new invention disclosures. As a result of IPP's efforts, over 100 NASA patent applications were filed and 93 patents awarded in FY 2007. Revenues realized from licenses of NASA-sponsored technologies exceeded $4 million in FY 2007.

- IPP completed six Centennial Challenge events and awarded $450,000 in combined prize money at two of them.

To complement the specialized centers and programs sponsored by the IPP, affiliated organizations and services have been formed to strengthen NASA's commitment to U.S. businesses and build upon NASA's experience in technology transfer.

Technology Infusion

Much of what we gain from our space exploration is in the scientific and technological progress that comes from the process of doing it. Many of those technologies

are the direct result of NASA-supported funding for both internal research and development (R&D) projects performed at NASA centers and external research from the small business community. As a result of these expanding needs for new capabilities to explore space, NASA missions often result in technologies which have applications beyond aerospace. These technologies, while targeted for integration into the mainstream NASA flight programs, can also be commercialized, creating new marketplace products and improving the quality of life for the American public right here on Earth.

For NASA, technology infusion is the process of strategically binding technical needs and potential solutions. These innovative solutions, be they hardware or software; enhancing or enabling; near-term or far-term; low technology readiness level (TRL) or high TRL; NASA internally or externally developed; must all be managed through some aspect of transition from their originating

source to the targeted challenges within NASA's programs and projects.

The IPP Technology Infusion element includes the Small Business Innovation Research (SBIR)/Small Business Technology Transfer (STTR) programs and the IPP Seed Fund programs. Together, these programs provide pathways from these originating sources to IPP's technology portfolio and provide enabling infrastructures that enhance the infusion of these technologies in NASA missions and programs. These programs allow the Agency to implement successful technology infusion and receive benefits in the following ways:

- Leverage limited program funds for technology development
- Leverage partners' funds/investments to achieve NASA's R&D goals
- Avoid additional program costs by providing a portfolio of technology solutions
- Accelerate technology maturation through concurrent R&D

SBIR projects often result in commercial products that benefit not only NASA, but also the economy, like this lightweight and flexible thin film solar cell battery charger that came out of work done at Glenn Research Center by PowerFilm Inc.

- Make informed decisions when selecting technologies for programs/projects/missions (i.e., better trade space information)
- Increase the return on its R&D investment with additional marketplace applications of technologies (benefits for both NASA and the public)

NASA issues annual program solicitations that set forth a substantial number of research topics and subtopic areas consistent with stated Agency needs or missions. Both the list of topics and the description of the topics and subtopics are sufficiently comprehensive to provide a wide range of opportunity for small business concerns to participate in NASA research or R&D programs. Topics and subtopics emphasize the need for proposals with advanced concepts to meet specific Agency research needs.

The NASA SBIR program <**http://www.sbir.nasa. gov**> provides seed money to small U.S. businesses to develop innovative concepts that meet NASA mission requirements. Each year, NASA invites small businesses to offer proposals in response to technical topics listed in the annual SBIR-STTR Program Solicitations. The NASA field centers negotiate and award the contracts, and then monitor the work.

NASA's SBIR program is implemented in three phases:

- **Phase I** is the opportunity to establish the feasibility and technical merit of a proposed innovation. Selected competitively, NASA Phase I contracts last 6 months and must remain under specific monetary limits.
- **Phase II** is the major research and development effort which continues the most promising of the Phase I projects based on scientific and technical merit, results of Phase I, expected value to NASA, company capability, and commercial potential. Phase II places greater emphasis on the commercial value of the innovation. The contracts are usually in effect for a period of 24 months and again must not exceed specified monetary limits.

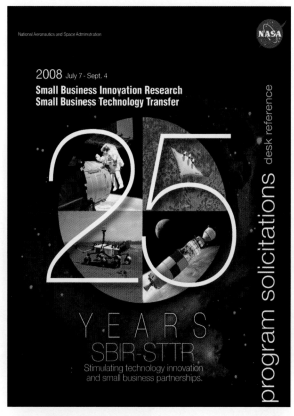

Small businesses develop technologies in response to specific NASA mission-driven needs, as presented in the NASA SBIR-STTR Program Solicitations.

- **Phase III** is the process of completing the development of a product to make it commercially available. While the financial resources needed must be obtained from sources other than the funding set aside for the SBIR, NASA may fund Phase III activities for follow-on development or for production of an innovation for its own use.

The NASA STTR program <**http://www.sbir.nasa. gov**> differs from the SBIR program in that the funding and technical scope is limited and participants must be teams of small businesses and research institutions that will conduct joint research.

SBIR/STTR Hallmarks of Success Videos are short videos about successful companies that have participated in the SBIR and STTR programs. Available online at <**http://sbir.nasa.gov/SBIR/successvideo.html**>, many of these videos also feature products that have been featured in the pages of *Spinoff*.

As another area of emphasis for technology infusion, the IPP established the Seed Fund with the following objectives that:

- Support NASA mission directorate program/project technology needs

- Provide "bridge" funding to centers in support of mission directorate programs

- Promote partnerships and cost sharing with mission directorate programs and industry

- Leverage resources with greater return on investment

Innovation Incubator

The IPP Innovation Incubator currently includes three distinct elements: Centennial Challenges, the FAST program, and Innovation Transfusion.

Centennial Challenges <**http://centennialchallenges. nasa.gov**> is NASA's program of prize contests to stimulate innovation and competition in solar system exploration and ongoing NASA mission areas. By making awards based on actual achievements, instead of proposals, Centennial Challenges seeks novel solutions to NASA's mission challenges from non-traditional sources of innovation in academia, industry, and the public. Challenges include:

- The **Beam Power Challenge**, which is designed to promote the development of new power distribution technologies. This initiative is managed by The Spaceward Foundation.

- The **Tether Challenge**, also managed by The Spaceward Foundation, whose purpose is to develop very strong tether material for use in various structural applications.

- The **Lunar Lander Challenge**, designed to accelerate technology developments supporting the commercial creation of a vehicle capable of ferrying cargo or humans back and forth between lunar orbit and the lunar surface. This project is run in conjunction with the X PRIZE Foundation.

- The **Astronaut Glove Challenge**, whose aim is to promote the development of glove joint technology, resulting in a highly dexterous and flexible glove that can be used by astronauts over long periods of time for space or planetary surface excursions. Volanz Aerospace Inc./Spaceflight America manages this program for NASA.

This year, researchers at Glenn Research Center were awarded their 100th "R&D 100" award, a designation from R&D Magazine that the Center has developed products that rank among the top 100 most technologically significant products of the year.

- The **Regolith Excavation Challenge** promotes the development of new technologies to excavate lunar regolith, which is a necessary first step toward lunar resource utilization. Sponsorship for this challenge is courtesy of the California Space Education and Workforce Institute (CSEWI).

- The **General Aviation Technology Challenge** is intended to bring about the development of new aviation technologies, which can improve the community acceptance, efficiency, door-to-door speed, and safety of future air vehicles. For this challenge, NASA has partnered with the Comparative Aircraft Flight Efficiency (CAFE) Foundation.

- The **Moon Regolith Oxygen (MoonROx) Challenge** is designed to promote the development of processes to extract oxygen from lunar regolith on the scale of a pilot plant. Like the Regolith Excavation Challenge, this contest is sponsored by the California Space Education and Workforce Institute (CSEWI).

Another aspect of the Innovation Incubator, the Facilitated Access to the Space Environment for Technology Development and Training (FAST) program, facilitates access to the space environment for testing by providing cost-shared access to parabolic flights.

A big challenge to technology infusion is the perceived risk by program and project managers, and they generally desire technologies to be at TRL 6 by their preliminary design review. A key element of achieving TRL 6 is demonstrating a technology in the relevant environment, including the gravity environment—from microgravity to lunar or Martian gravity levels. Toward that end, the FAST program seeks to provide more opportunities for advancing TRLs by providing partnership opportunities to demonstrate technologies in these environments.

Currently, space technology development often stalls at the mid-technology readiness levels due to lack of opportunities to test prototypes in relevant environments. In addition, limited testing opportunities often have high

The Beam Power Challenge is designed to promote the development of new power distribution technologies, which can then be applied to many aspects of space exploration, including surface- or space-based point-to-point power transmission or delivery for robotic and/or human expeditions to planetary surfaces.

associated costs or require lengthy waits. NASA recently selected a commercial service provider for parabolic aircraft flight to simulate multiple gravity environments. FAST will purchase services through this new procurement mechanism and provide partnership opportunities aimed at advancing needed space technologies to higher technology readiness levels. The objective is to provide advanced technologies with risk levels that enable more infusion, meeting the priorities of NASA's mission directorates and their programs and projects.

Just as NASA is able to provide support to outside research agencies, there are many creative external organizations implementing innovative processes and methods that could be of benefit to NASA. Innovation

Transfusion involves reaching out to some of these organizations on an ad hoc basis. Innovation Transfusion will be a focused activity to more strategically realize the potential from external creativity. Innovation Transfusion will accomplish this goal through the implementation of NASA Innovation Ambassadors, a technical training program that places NASA technical employees at external organizations, and NASA Innovation Scouts, NASA teams participating in focused workshops to exchange information on specific innovations with external organizations. Innovation Transfusion will follow an annual process to select ambassadors and schedule scout workshops with the goal of incorporating innovations into NASA to meet Agency needs. Innovation Transfusion is expected to place its first Innovation Ambassador at an external organization in 2008.

Partnership Development

The National Aeronautics and Space Act of 1958 and subsequent legislation recognize the transfer of federally owned or originated technology to be a national priority. Accordingly, NASA is obliged to provide for the widest practicable dissemination of information concerning results of NASA's activities. The legislation specifically mandates that each Federal agency have a formal technology transfer program, and take an active role in transferring technology to the private sector and state and local governments for the purposes of commercial and other application of the technology for the national benefit. In accordance with NASA's obligations under mandating legislation, IPP, on behalf of NASA, facilitates the transfer of technology to which NASA has title for commercial application and other national benefit. IPP seeks out potential licensees and negotiates license agreements to transfer NASA technology. IPP typically facilitates over 50 new licenses with the private sector each year.

Over the past 50 years, NASA has performed some amazing feats and made many powerful discoveries. However, aware that it does not have all the answers, NASA continuously looks for partners to help develop new technologies, inform the public about advancements, and to take existing business practices to a new level of effectiveness.

To complement IPP's Partnership Development efforts, the **National Technology Transfer Center (NTTC)** <**http://www.nttc.edu**> links U.S. industry with Federal laboratories and universities that have the technologies, the facilities, and the world-class researchers that industry needs to maximize product development opportunities. The NTTC has worked with NASA since 1989, providing the services and capabilities needed to meet the changing needs of NASA for managing intellectual property and creating technology partnerships.

The **Federal Laboratory Consortium for Technology Transfer (FLC)** <**http://www.federallabs.org**> was organized in 1974 and formally chartered by the Federal Technology Transfer Act of 1986 to promote and strengthen technology transfer nationwide. More than 700 major Federal laboratories and centers, including NASA, are currently members. The mission of the FLC is twofold: 1) To promote and facilitate the rapid movement of Federal laboratory research results and technologies into the mainstream U.S. economy by fostering partnerships and collaboration with the private sector, academia, economic development organizations, and other entities engaged in technology development; and 2) To use a coordinated program that meets the technology transfer support needs of FLC member laboratories, agencies, and their potential partners in the transfer process.

Affiliate Organizations

The road to technology commercialization begins with the basic and applied research results from the work of scientists, engineers, and other technical and management personnel. The **NASA Scientific and Technical Information (STI)** program <**http://www.sti.nasa.gov**> provides wide dissemination of NASA's research results.

The NASA STI program offers users Internet access to its database of over 4 million citations, as well as many in full text; online ordering of documents; and the NASA STI Help Desk (help@sti.nasa.gov) for assistance in accessing STI resources and information. Free registration

The Innovative Partnerships Program brokers many types of agreements between the Agency and industry, like the Space Act Agreement being signed here, where Goddard Space Flight Center and Northrop Grumman are partnering to answer key questions in climate change and planetary science.

Each year, the Innovative Partnerships Program hosts the New Technology Reporting (NTR) Program to recognize innovators who have filed NTRs, received patents, or made significant efforts to assist in establishing partnerships or transferring technology. In May 2007, Greg Olsen, entrepreneur and the third private citizen to orbit the Earth aboard the International Space Station, spoke at the event about how his company benefited from involvement with the NASA SBIR program.

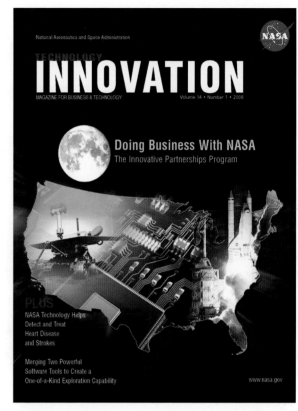

Technology Innovation is one of NASA's magazines for business and technology, published by the Innovative Partnerships Program.

with the program is available for qualified users through the NASA Center for AeroSpace Information.

The **NASA Technology Portal** <**http://technology. nasa.gov**> provides access to NASA's technology inventory and numerous examples of the successful transfer of NASA-sponsored technology. TechFinder, the main feature of the Internet site, allows users to search technologies and success stories, as well as submit requests for additional information.

Working closely with NASA's IPP offices and public information offices, the Space Foundation each year recognizes individuals, organizations, and companies that develop innovative products based on space technology. These honorees are enshrined in the **Space Technology Hall of Fame** <**http://www.spacetechhalloffame.org**>. The ever-growing list of inductees showcases the significant contributions that space technology has made to improve the quality of life for everyone around the world.

For more than three decades, **NASA *Tech Briefs*** <**http://www.techbriefs.com**> has reported to industry on any new, commercially significant technologies developed in the course of NASA R&D efforts. The monthly magazine features innovations from NASA, industry partners, and contractors that can be applied to develop new or improved products and solve engineering or manufacturing problems.

Authored by the engineers or scientists who performed the original work, the briefs cover a variety of disciplines, including computer software, mechanics, and life sciences. Most briefs offer a free supplemental technical support package, which explains the technology in greater detail and provides contact points for questions or licensing discussions.

Technology Innovation <**http://www.ipp.nasa. gov/innovation**> is published quarterly by the Innovative Partnerships Program. Regular features include current news and opportunities in technology transfer and commercialization, and innovative research and development.

NASA *Spinoff* <**http://www.sti.nasa.gov/tto**> is an annual print and electronic publication featuring successful commercial and industrial applications of NASA technology, current research and development efforts, and the latest developments from the NASA Innovative Partnerships Program. ❖

NASA Innovative Partnerships Program Network Directory

T he 2008 NASA Innovative Partnerships Program (IPP) network extends from coast to coast. For specific information concerning technology partnering activities, contact the appropriate personnel at the facilities listed or visit the Web site: **<http://www.ipp.nasa.gov>**. General inquiries may be forwarded to the Spinoff Program Office at **<spinoff@sti.nasa.gov>**.

To publish a story about a product or service you have commercialized using NASA technology, assistance, or know-how, contact the NASA Center for AeroSpace Information, or visit: **<http://www.sti.nasa. gov/tto/contributor.html>**.

★ **NASA Headquarters** manages the Spinoff Program.

▲ **Innovative Partnerships Program Offices** at each of NASA's 10 field centers represent NASA's technology sources and manage center participation in technology transfer activities.

■ **Allied Organizations** support NASA's IPP objectives.

NASA HQ — NASA Headquarters
ARC — Ames Research Center
CASI — NASA Center for AeroSpace Information
DFRC — Dryden Flight Research Center
FLC — Federal Laboratory Consortium
GRC — Glenn Research Center
GSFC — Goddard Space Flight Center
JPL — Jet Propulsion Laboratory
JSC — Johnson Space Center
KSC — Kennedy Space Center
LaRC — Langley Research Center
MSFC — Marshall Space Flight Center
NTTC — National Technology Transfer Center
SF — Space Foundation
SPC/Fuentek — Systems Planning Corporation/Fuentek
SSC — Stennis Space Center
Tech Briefs — Tech Briefs Media Group

Centennial Challenge Allied Organizations

1. Spaceward Foundation
2. X PRIZE Foundation
3. Volanz Aerospace Inc./Spaceflight America
4. California Space Education and Workforce Institute
5. Comparative Aircraft Flight Efficiency Foundation

★ NASA Headquarters

National Aeronautics and Space Administration
300 E Street, SW
Washington, DC 20546
NASA *Spinoff* Publication Manager:
Janelle Turner
Phone: (202) 358-0704
E-mail: janelle.b.turner@nasa.gov

▲ Field Centers

Ames Research Center
National Aeronautics and Space Administration
Moffett Field, California 94035
Chief, Technology Partnerships Division:
Lisa Lockyer
Phone: (650) 604-1754
E-mail: lisa.l.lockyer@nasa.gov

Dryden Flight Research Center
National Aeronautics and Space Administration
4800 Lilly Drive, Building 4839
Edwards, California 93523-0273
Chief, Technology Transfer Partnerships Office:
Greg Poteat
Phone: (661) 276-3872
E-mail: gregory.a.poteat@nasa.gov

Glenn Research Center
National Aeronautics and Space Administration
21000 Brookpark Road
Cleveland, Ohio 44135
Director, Technology Transfer and Partnerships Office:
Kathleen Needham
Phone: (216) 433-2802
E-mail: kathleen.k.needham@nasa.gov

Goddard Space Flight Center
National Aeronautics and Space Administration
Greenbelt, Maryland 20771
Chief, Innovative Partnerships Program Office:
Nona K. Cheeks
Phone: (301) 286-5810
E-mail: nona.k.cheeks@nasa.gov

Jet Propulsion Laboratory
National Aeronautics and Space Administration
4800 Oak Grove Drive
Pasadena, California 91109
Manager, Commercial Program Office:
Andrew Gray
Phone: (818) 354-4906
E-mail: andrew.a.gray@nasa.gov

Johnson Space Center
National Aeronautics and Space Administration
Houston, Texas 77058
Director, Technology Transfer Office:
Michele Brekke
Phone: (281) 483-4614
E-mail: michele.brekke@nasa.gov

Kennedy Space Center
National Aeronautics and Space Administration
Kennedy Space Center, Florida 32899
Chief, Technology Programs and Partnerships Branch:
David R. Makufka
Phone: (321) 867-6227
E-mail: david.r.makufka@nasa.gov

Langley Research Center
National Aeronautics and Space Administration
Hampton, Virginia 23681-2199
Manager, Advanced Planning and Partnerships Office:
Brian Beaton
Phone: (757) 864-2192
E-mail: brian.f.beaton@nasa.gov

Marshall Space Flight Center
National Aeronautics and Space Administration
Marshall Space Flight Center, Alabama 35812
Acting Director, Technology Transfer Office:
James Dowdy
Phone: (256) 544-7604
E-mail: jim.dowdy@nasa.gov

Stennis Space Center
National Aeronautics and Space Administration
Stennis Space Center, Mississippi 39529
Manager, Innovative Partnerships Program Office:
Ramona Pelletier Travis
Phone: (228) 688-3832
E-mail: ramona.e.travis@ssc.nasa.gov

■ Allied Organizations

National Technology Transfer Center (NTTC)
Wheeling Jesuit University
Wheeling, West Virginia 26003
Darwin Molnar, Vice President
Phone: (800) 678-6882
E-mail: dmolnar@nttc.edu

Systems Planning Corporation
1000 Wilson Blvd,
Arlington, Virginia 22209
James M. Kudla, Vice President, Corporate Communications
Phone: (703) 351-8238
E-mail: jkudla@sysplan.com

Fuentek LLC
85 Goldfinch Lane,
Apex, North Carolina 27523
Laura A. Schoppe, Director
Phone: (919) 303-5874
E-mail: laschoppe@fuentek.com

Space Foundation
310 S. 14th Street
Colorado Springs, Colorado 80904
Kevin Cook, Director, Space Technology Awareness
Phone: (719) 576-8000
E-mail: kevin@spacefoundation.org

Federal Laboratory Consortium
300 E Street, SW
Washington, DC 20546
John Emond, Collaboration Program Manager
Phone: (202) 358-1686
E-mail: john.l.emond@nasa.gov

Tech Briefs Media Group
1466 Broadway, Suite 910
New York, NY 10036
Joseph T. Pramberger, Publisher
(212)490-3999
www.techbriefs.com

NASA Center for AeroSpace Information
Spinoff Project Office
7115 Standard Drive
Hanover, Maryland 21076-1320
E-mail: spinoff@sti.nasa.gov

Daniel Lockney, Editor/Writer
Phone: (301) 621-0224
E-mail: dlockney@sti.nasa.gov

Andréa Miralia, Editor/Writer

Gareth Williams, Editor/Writer

John Jones, Graphic Designer

Deborah Drumheller, Publications Specialist

NASA Technologies Benefiting Society

Health and Medicine

1. Robotics Offer Newfound Surgical Capabilities (MA)
2. In-Line Filtration Improves Hygiene and Reduces Expense (MN)
3. LED Device Illuminates New Path to Healing (WI)
4. Polymer Coats Leads on Implantable Medical Device (MN)
5. Lockable Knee Brace Speeds Rehabilitation (AR)
6. Robotic Joints Support Horses and Humans (CT)
7. Photorefraction Screens Millions for Vision Disorders (AL)
8. Periodontal Probe Improves Exams, Alleviates Pain (VA)
9. Magnetic Separator Enhances Treatment Possibilities (IN)

Transportation

10. Lithium Battery Power Delivers Electric Vehicles to Market (NV)
11. Advanced Control System Increases Helicopter Safety (CA)
12. Aerodynamics Research Revolutionizes Truck Design (CO)
13. Engineering Models Ease and Speed Prototyping (MI)
14. Software Performs Complex Design Analysis (ID)

Public Safety

15. Space Suit Technologies Protect Deep-Sea Divers (AZ)
16. Fiber Optic Sensing Monitors Strain and Reduces Costs (VA)
17. Polymer Fabric Protects Firefighters, Military, and Civilians (SC)
18. Advanced X-Ray Sources Ensure Safe Environments (CA)

Consumer, Home, and Recreation

19. Wireless Fluid-Level Measurement System Equips Boat Owners (VA)
20. Mars Cameras Make Panoramic Photography a Snap (TX)
21. Experiments Advance Gardening at Home and in Space (CO)
22. Space Age Swimsuit Reduces Drag, Breaks Records (CA)
23. Immersive Photography Renders 360° Views (TN)
24. Historic Partnership Captures Our Imagination (NJ)
25. Outboard Motor Maximizes Power and Dependability (WI)
26. Space Research Fortifies Nutrition Worldwide (MD)
27. Aerogels Insulate Missions and Consumer Products (MA)

Environmental and Agricultural Resources

28. Computer Model Locates Environmental Hazards (CA)
29. Battery Technology Stores Clean Energy (CA)
30. Robots Explore the Farthest Reaches of Earth and Space (CA)
31. Portable Nanomesh Creates Safer Drinking Water (VT)
32. Innovative Stemless Valve Eliminates Emissions (WY)
33. Web-Based Mapping Puts the World at Your Fingertips (MS)

Computer Technology

34. Program Assists Satellite Designers (MD)
35. Water-Based Coating Simplifies Circuit Board Manufacturing (OH)
36. Software Schedules Missions, Aids Project Management (TX)
37. Software Analyzes Complex Systems in Real Time (CA)
38. Wireless Sensor Network Handles Image Data (CO)
39. Virtual Reality System Offers a Wide Perspective (MD)
40. Software Simulates Sight: Flat Panel Mura Detection (WA)
41. Inductive System Monitors Tasks (ME)
42. Mars Mapping Technology Brings Main Street to Life (CA)
43. Intelligent Memory Module Overcomes Harsh Environments (TX)
44. Integrated Circuit Chip Improves Network Efficiency (MD)

Industrial Productivity

45. Novel Process Revolutionizes Welding Industry (MI)
46. Sensors Increase Productivity in Harsh Environments (CA)
47. Portable Device Analyzes Rocks and Minerals (CA)
48. NASA Design Strengthens Welds (MN)
49. Polyimide Boosts High-Temperature Performance (OH)
50. NASA Innovation Builds Better Nanotubes (TX)

Locations of the companies featured in *Spinoff* 2008 are indicated by the corresponding numbers on this map.

The Nation's investment in NASA's aerospace research has brought practical benefits back to Earth in the form of commercial products and services in the fields of health and medicine; transportation; public safety; consumer, home, and recreation goods; environmental and agricultural resources; computer technology; and industrial productivity. *Spinoff*, NASA's premier annual publication, features these commercialized technologies. Since its inception in 1976, *Spinoff* has profiled more than 1,600 NASA-derived products from companies across the Nation. An online archive of all stories from the first issue of *Spinoff* to the latest is available in the *Spinoff* database at www.sti.nasa.gov/spinoff/database.

Visit the Innovative Partnerships Program at **http://www.ipp.nasa.gov.**